Lecture Notes in Networks and Systems

Volume 249

The series "Lecture Notes in Networks and Systems" publishes the latest developments in Networks and Systems—quickly, informally and with high quality. Original research reported in proceedings and post-proceedings represents the core of LNNS.

Volumes published in LNNS embrace all aspects and subfields of, as well as new challenges in, Networks and Systems.

The series contains proceedings and edited volumes in systems and networks, spanning the areas of Cyber-Physical Systems, Autonomous Systems, Sensor Networks, Control Systems, Energy Systems, Automotive Systems, Biological Systems, Vehicular Networking and Connected Vehicles, Aerospace Systems, Automation, Manufacturing, Smart Grids, Nonlinear Systems, Power Systems, Robotics, Social Systems, Economic Systems and other. Of particular value to both the contributors and the readership are the short publication timeframe and the world-wide distribution and exposure which enable both a wide and rapid dissemination of research output.

The series covers the theory, applications, and perspectives on the state of the art and future developments relevant to systems and networks, decision making, control, complex processes and related areas, as embedded in the fields of interdisciplinary and applied sciences, engineering, computer science, physics, economics, social, and life sciences, as well as the paradigms and methodologies behind them.

Indexed by SCOPUS, INSPEC, WTI Frankfurt eG, zbMATH, SCImago.

All books published in the series are submitted for consideration in Web of Science.

More information about this series at https://link.springer.com/bookseries/15179

Luigi Troiano · Alfredo Vaccaro ·
Roberto Tagliaferri · Nishtha Kesswani ·
Irene Díaz Rodriguez · Imene Brigui ·
Domenico Parente
Editors

Advances in Deep Learning, Artificial Intelligence and Robotics

Proceedings of the 2nd International Conference on Deep Learning, Artificial Intelligence and Robotics, (ICDLAIR) 2020

Springer

Editors
Luigi Troiano
Department of Management and Innovation Systems
University of Salerno
Fisciano, Italy

Roberto Tagliaferri
Department of Management and Innovation Systems
University of Salerno
Fisciano, Italy

Irene Díaz Rodriguez
Department of Computer Science
University of Oviedo
Gijón, Spain

Domenico Parente
Department of Management and Innovation Systems
University of Salerno
Fisciano, Italy

Alfredo Vaccaro
Department of Engineering
University of Sannio
Benevento, Italy

Nishtha Kesswani
Department of Computer Science
Central University of Rajasthan
Ajmer, Rajasthan, India

Imene Brigui
Emlyon Business School
Écully, France

ISSN 2367-3370 ISSN 2367-3389 (electronic)
Lecture Notes in Networks and Systems
ISBN 978-3-030-85364-8 ISBN 978-3-030-85365-5 (eBook)
https://doi.org/10.1007/978-3-030-85365-5

This Springer imprint is published by the registered company Springer Nature Switzerland AG
The registered company address is: Gewerbestrasse 11, 6330 Cham, Switzerland

Organization

Program Chairs

Roberto Tagliaferri	DISA-MIS, University of Salerno, Italy
Alfredo Vaccaro	Università del Sannio, Italy
Rajiv Misra	IIT, India
Irene Díaz Rodriguez	University of Oviedo
Luigi Troiano	University of Salerno, Italy
Imene Brigui	Emlyon Business School
Nishtha Kesswani	Central University of Rajasthan, India

Program Committee

Roberto Tagliaferri	DISA-MIS, University of Salerno, Italy
Alfredo Vaccaro	Università del Sannio, Italy
Irene Díaz Rodriguez	University of Oviedo
Marie-Hélène Abel	Heudiasyc, Université de Technologie de Compiègne
Luigi Troiano	University of Salerno, Italy
Imene Brigui	Emlyon Business School
Nishtha Kesswani	Central University of Rajasthan, India

Preface

The 2nd International Conference on Deep Learning, Artificial Intelligence and Robotics (ICDLAIR 2020) was organized by University of Salerno, Italy, in collaboration with International Association of Academicians (IAASSE), USA, and held in December 2020.

Due to the outbreak of COVID-19, this year's conference was organized as fully virtual conference. This was an incredible opportunity to experiment with a conference format that we plan to continue in the future. In order to prepare ICDLAIR 2020, the organizing committees, reviewers, session chairs as well as all the authors and presenters have made a lot of efforts and contributions. Thank you very much for always being supportive to the conference. We could not have pulled off this convention without all of your hard work and dedication.

This proceedings book includes the results from the conference. The scope of this conference includes all subareas of AI, with broad coverage of traditional topics like robotics, statistical learning and deep learning techniques. However, the organizing committee expressly encouraged work on the applications of DL and AI in the important fields of engineering, including computer electronics, electrical, mechanical, chemical, textile, together with health care and agriculture, business and social media and other relevant domains.

The conference welcomed papers on the following (but not limited to) research topics:

- Deep Learning: Applications of deep learning in various engineering streams, neural information processing systems, training schemes, GPU computation and paradigms, human–computer interaction, genetic algorithm, reinforcement learning, natural language processing, social computing, user customization, embedded computation, automotive design and bioinformatics.
- Artificial Intelligence: Automatic control, natural language processing, data mining and machine learning tools, fuzzy logic, heuristic optimization techniques (membrane-based separation, wastewater treatment, process control, etc.) and soft computing.

- Robotics: Automation and advanced control-based applications in engineering, neural networks on low powered devices, human–robot interaction and communication, cognitive, developmental and evolutionary robotics, fault diagnosis, virtual reality, space and underwater robotics, simulation and modeling, bioinspired robotics, cable robots, cognitive robotics, collaborative robotics, collective and social robots and humanoid robots.

It was a collaborative platform for academic experts, researchers and corporate professionals for interacting their research in various domain of engineering like robotics, data acquisition, human–computer interaction, genetic algorithm, sentiment analysis as well as usage of AI and advanced computation in various industrial challenges-based applications such as user customization, augmented reality, voice assistants, reactor design, product formulation/synthesis, embedded system design, membrane-based separation for protecting environment along with wastewater treatment, rheological properties estimation for Newtonian and non-Newtonian fluids used in micro-processing industries and fault detection.

ICDLAIR 2020 aims to become premier international conference for in-depth discussion on the most-up-to-date and innovative ideas, research projects and practices in the field of artificial intelligence with application to multiple fields, robotics included. Today, the world develops with unprecedented speed, which is only possible with advances in various fields of sciences and engineering. We are very pleased to contribute to this important global effort. We sincerely hope that the academic community and industrial practitioners will continue to support us in our attempts to provide even more meaningful conferences with numerous critical idea exchanges, diverse opportunities for fruitful networking, future collaborations and international cooperation.

Fisciano, Italy Luigi Troiano
Benevento, Italy Alfredo Vaccaro
Fisciano, Italy Roberto Tagliaferri
Ajmer, India Nishtha Kesswani
Gijón, Spain Irene Díaz Rodriguez
Écully, France Imene Brigui
Fisciano, Italy Domenico Parente
January 2021

Contents

Deep Learning Based Classification System for Recognizing Local Spinach 1
Mirajul Islam, Nushrat Jahan Ria, Jannatul Ferdous Ani, Abu Kaisar Mohammad Masum, Sheikh Abujar, and Syed Akhter Hossain

Fusion of Deep Learning Features for Wild Animal Detection 15
Anamika Dhillon and Gyanendra K. Verma

Forecast Extreme CPU Usages Under Peak Load Using Envelop EM Semi-supervised Learning 25
Nitin Khosla and Dharmendra Sharma

Classifying Case Facts and Predicting Legal Decisions of The Indian Central Information Commission: a Natural Language Processing Approach 35
Arvind Kalia, Naveen Kumar, and Nischay Namdev

On the Implementation of a Semantic Model for Intelligent Vehicle Navigation 47
Alessandro Correa Victorino and Marie-Hélène Abel

Twenty Years of Assessment of Drought Conditions Over The Gangapur Region of Aurangabad District in Maharashtra (India) Using Remote Sensing Indices 55
Rashmi R. Nitwane, Vaishali D. Bhagile, and R. R. Deshmukh

Educational Resources Recommender System for Teachers: Why and How? 71
Nader N. Nashed, Christine Lahoud, and Marie-Hélène Abel

Hybrid Cloud Native Framework for Smart City Applications 81
Amit Kar

A Two-Stage Intrusion Detection System (TIDS) for Internet of Things 89
Monika Vishwakarma and Nishtha Kesswani

Breast Cancer Detection Using Deep Convolutionnal Neural Networks and Raspberry Pi 3 Board 99
Iman Elawady, Abdel Kamal Bensaada, and El Mostafa Litim

An Improved Approach for Removal of Salt and Pepper Noise in MR Images 111
Monika Jyotiyana, Nishtha Kesswani, and Anupam Agarwal

Dual-Pathway Deep CNN for Offline Writer Identification 119
Naresh Purohit and Subhash Panwar

A Multi-Agent System for Industrial Simulators Design 129
Kenza Redjimi and Mohammed Redjimi

Comparison of Current Convolutional Neural Network Architectures for Classification of Damaged and Undamaged Cars 141
Yavuz Ünal, Şaban Öztürk, Muhammed Nuri Dudak, and Mehmet Ekici

Intelligent Agents for Multi-user Preference Elicitation 151
Ons Nefla, Iméne Brigui, Meltem Ozturk, and Paolo Viappiani

Automatic Grading of Knee Osteoarthritis on the Kellgren-Lawrence Scale from Radiographs Using Convolutional Neural Networks 163
Sudeep Kondal, Viraj Kulkarni, Ashrika Gaikwad, Amit Kharat, and Aniruddha Pant

Robust Model for Rural Education Using Deep Learning and Robotics 175
Sailesh Iyer

Action Recognition for Anomaly Detection Using Transfer Learning and Deep Architectures 183
Francisco Luque Sánchez, Siham Tabik, and Francisco Herrera

Personalised Federated Learning with BERT Fine Tuning. Case Study on Twitter Sentiment Analysis 193
Jose Antonio Ruiz-Millán, Eugenio Martínez-Cámara, M. Victoria Luzón, and Francisco Herrera

Use of Artificial Intelligence in Human Resource Management: "Application of Machine Learning Algorithms to an Intelligent Recruitment System" 203
Said Achchab and Yassine Khallouk Temsamani

Combining Active Semi-supervised Learning and Rare Category Detection 217
Rohan Loveland and Noah Kaplan

Author Index 231

Deep Learning Based Classification System for Recognizing Local Spinach

Mirajul Islam, Nushrat Jahan Ria, Jannatul Ferdous Ani, Abu Kaisar Mohammad Masum, Sheikh Abujar, and Syed Akhter Hossain

Abstract A deep learning model gives an incredible result for image processing by studying from the trained dataset. Spinach is a leaf vegetable that contains vitamins and nutrients. In our research, a Deep learning method has been used that can automatically identify spinach and this method has a dataset of a total of five species of spinach that contains 3785 images. Four Convolutional Neural Network (CNN) models were used to classify our spinach. These models give more accurate results for image classification. Before applying these models there is some preprocessing of the image data. For the preprocessing of data, some methods need to happen. Those are RGB conversion, filtering, resize and rescaling, and categorization. After applying these methods image data are preprocessed and ready to be used in the classifier algorithms. The accuracy of these classifiers is in between 98.68 and 99.79%. Among those models, VGG16 achieved the highest accuracy of 99.79%.

M. Islam (✉) · N. J. Ria · J. F. Ani · A. K. M. Masum
Department of Computer Science and Engineering, Daffodil International University, Dhaka 1209, Bangladesh
e-mail: merajul15-9627@diu.edu.bd

N. J. Ria
e-mail: nushrat15-9771@diu.edu.bd

J. F. Ani
e-mail: jannatul15-10483@diu.edu.bd

A. K. M. Masum
e-mail: mohammad15-6759@diu.edu.bd

S. Abujar
Department of Computer Science and Engineering, Independent University Bangladesh, Dhaka 1229, Bangladesh
e-mail: sheikh.csesets@iub.edu.bd

S. A. Hossain
Department of Computer Science and Engineering, University of Liberal Arts Bangladesh, Dhaka 1207, Bangladesh
e-mail: akhter.hossain@ulab.edu.bd

L. Troiano et al. (eds.), *Advances in Deep Learning, Artificial Intelligence and Robotics*, Lecture Notes in Networks and Systems 249,
https://doi.org/10.1007/978-3-030-85365-5_1

Keywords Spinach recognition · Convolutional neural network (CNN) · Deep learning · Image classification · Evaluation metric

1 Introduction

Spinach is one of the favorable foods for the human body. It strengthens the bones of our body. It produces vitamin-A, vitamin-C, vitamin-E, vitamin-K, potassium, magnesium, iron, calcium, copper, phosphorous, zinc, selenium, folate, betaine, folic acid, protein, niacin, omega-3 fatty acids, and many more important nutrient [1]. It greatly enhances our ability to repel diseases. Also, some species are very important for our hair, eyes, heart, kidney, and skin. Spinach is a very popular food in the rural areas of Bangladesh. There are 107 types of spinach available in Bangladesh [2]. Although many of these are available in the market of rural areas, only a few species are found in urban areas. The spinach looks quite similar so most of the people and children of the new generation usually cannot get to recognize them. Even they do not know the name of these spinach. As a solution, we used several Convolution Neural Networks (CNN) to automatically classify different species of spinach based on spinach's leaves. We used five species of spinach for this research work. These are Jute spinach (*Corchorus olitorius*) which is locally known as Pat Shak, Malabar spinach (*Basella alba*) which is locally known as Pui Shak, Red spinach (*Amaranthus gangeticus*) which is locally known as Lal Shak, Taro spinach (*Colocasia esculenta*) which is locally known as Kochu Shak and Water spinach (*Ipomoea aquatica*) which is locally called as Kolmi Shak [2]. We preprocessed the data then we trained four different models for this classification. We used InceptionV3, Xception, VGG19, and VGG16 to identify those spinach.

CNN based image classification performs a vast advancement. There are two types of image classification, one is supervised and another one is unsupervised. Deep neural networks play the most important role in image analysis. Image classification is the primary domain, in which deep neural networks are being used. The model takes the given input images and provides classified output images for identifying whether the image represents a specific class or not. CNN model that performs perfect image classification accuracy for different images. Inception V3 architecture is one of the best models for image data analysis. Human performance with efficiency has been accomplished by Inception V3. The CNN architecture uses three-dimensional convolutions to read different input images. CNN's are the recent state-of-the-art methods in image classification. Natural images are learned by the CNNs, showing strong performance and encountering the accuracy of human expert systems. Finally, these statements conclude that CNNs can classify different images into different classes more accurately.

The remainder of this paper is organized as follows: Sect. 2 describes some of the work that is similar to our work. Section 3 describes our dataset, pre-processing steps, and the entire process. In Sect. 4 shows the results and graphs for each model and compares each of them. And in Sect. 5 describes our work summary and future work.

2 Related Work

In this sector of leaf recognition and classification, we studied some related research work. Wu et al. [3] proposed a Probabilistic Neural Network (PNN) with data processing techniques and images. At first, they captured digital leaf images. Then they processed those images. They extracted the features from the images. The features are orthogonalized by Principal Components Analysis (PCA). After that, they trained the data in Probabilistic Neural Network (PNN) and then tested the data in Probabilistic Neural Network (PNN). Finally, they display and compare the results. Their proposed algorithm shows 90% accuracy for the model. Wang et al. [4] proposed a deep learning method to automatically and accurately estimate disease severity security, disease management, and yield loss prediction for diagnosing plant disease severity. This method used an image-based plant disease recognition, this work proposes deep learning models for image-based automatic diagnosis of plant disease severity. They used the VGG16 model for this method because this model gave a better performance than other models. VGG16 model achieves an accuracy of 90.4%. Sun et al. [5] proposed a recognition plant model by using the BJFU100 dataset and deep learning model. This model contains 10,000 images of 100 ornamental plant species. BJFU100 has collected data from natural scenes by mobile devices. The Deep Residual Network method has been used to automatically identify the required representations of the classification. There is a 26 layer deep learning model consisting of 8 residual building blocks designed for uncontrolled plant identification. The proposed model recognition accuracy is 91.78% on the BJFU100 dataset. In recent years, Zhu et al. [6] proposed a system which is the possibility of the freshness identification of Spinach preserved at different temperatures by using Hyperspectral Imaging. They have applied three models. But there are ELM models performed best and the accuracy was achieved 100% of ELM models for the two spectral systems in the freshness detection of spinach leaves preserved at 4 °C and 20 °C. In this paper, Yalcin and Razavi [7] are mainly using a Convolutional Neural Network (CNN) model for classifying different plant species and identifying different plant species. Their proposed CNN architecture can automatically classify images of sixteen kinds of plants and achieve an accuracy of 97.47%. Kaya et al. [8] in plant classification models (2019), machine learning algorithms Deep Neural Networks (DNNs) have been applied to different data sets. They had designed and implemented five classification models including the baseline model (end-to-end CNN model) by applying four transfer learning strategies on deep learning-based models. Priya et al. [9] implemented plant leaf recognition through the kNN method. Firstly, Classifier tested with a Flavia dataset and a real dataset then compared with the KNN approach of plant leaf recognition. KNN approach had produced high accuracy and less execution time.

3 Research Methodology

In our work at first, we prepare our dataset with raw images, then we have divided those images into two portions, one is the training set and another is the test set, shown in Table 1. The five types of spinach are divided separately in both training and test sets (Fig. 1). Then the images from both sets are prepared for input into the model through pre-processing, shown in Fig. 2. Then four CNN models were trained with the data from our training set to classify the spinach. We used four CNN models to classify our data and evaluate their performances. The accuracy of the model has been measured on the ability to classify the data in the test set. These four models are InceptionV3 [10, 11], Xception [12], VGG19 [13], and VGG16 [13].

3.1 Collection and Properties of Dataset

We have collected a total of 3785 images of five types of fresh spinach leaves. These images have been collected from the local vegetable market and spinach fields. All images are captured in a fixed background. The images have been divided into two parts. The training set contains 80% of the images and the test set which contains 20% of images for testing the model, shown in Table 1. From the data distribution, it is a stabilized dataset that we can convey.

3.2 Data Pre-processing

Data pre-processing is a common method to process the data for implementing the model. In pre-processing, we take the raw images then apply different techniques to prepare the input data for the next stage. The main purpose of this image pre-processing is to improve the quality of the information of the data so that the machine can understand and analyze those data easily. In this data Pre-processing process,

Table 1 Distribution data into training and test set and the relation between spinach category and labels

Category	Total image	For training set	For test set	Target label
Jute spinach	750	600	150	0
Malabar spinach	758	606	152	1
Red spinach	761	609	152	2
Taro spinach	772	618	154	3
Water spinach	744	595	149	4
	3785	3028	757	

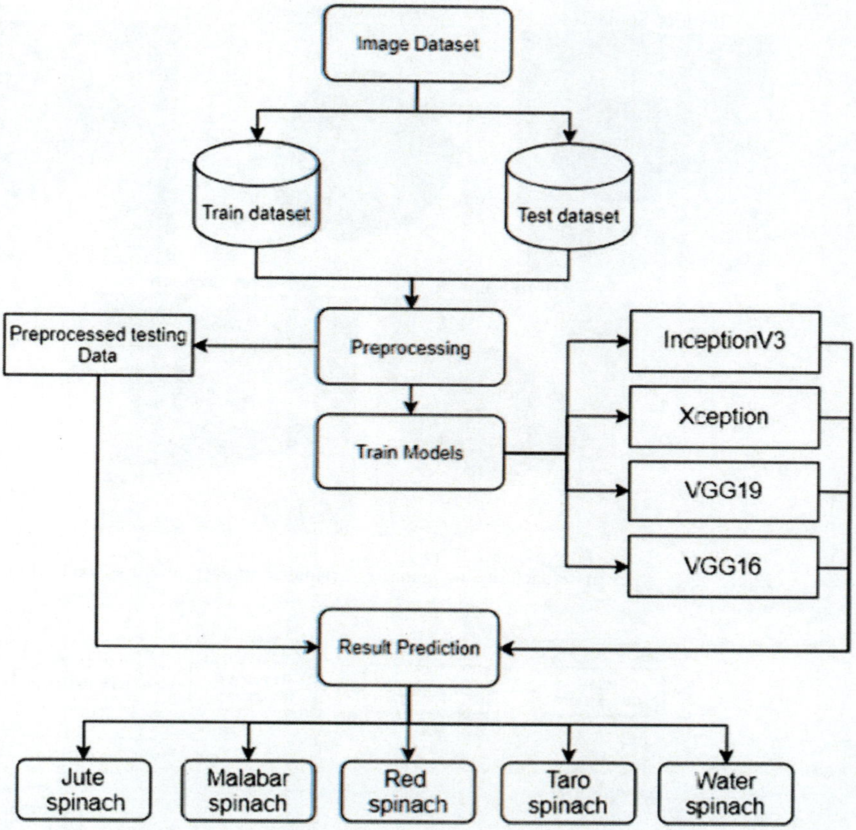

Fig. 1 Classification procedure

the images have been converted into RGB images. Then we filter or decrease the noise of the raw images. After that resize and rescale those data to process and then categorize the input data. Then the data is preprocessed and ready to train in the model. After preparing our dataset, to input, all the images into the model all images have been resized into height 224, width 224, and RGB color channel 3. Because our dataset contains different sizes of the images. $224 \times 224 \times 3$ is the input shape of our model (Fig. 2).

Then performing an image data generator, which is the main part of image preprocessing (Fig. 3). At first, all the images from the training and test set have been rescaled the values between 0 and 1. This process is known as Min–Max Normalization (MMN) [14]. Usually, the value of an RGB image is between 0 and 255. The equation for MMN as follows.

Fig. 2 Five types of spinach. **a** Jute spinach, **b** malabar spinach, **c** red spinach, **d** taro spinach, **e** water spinach

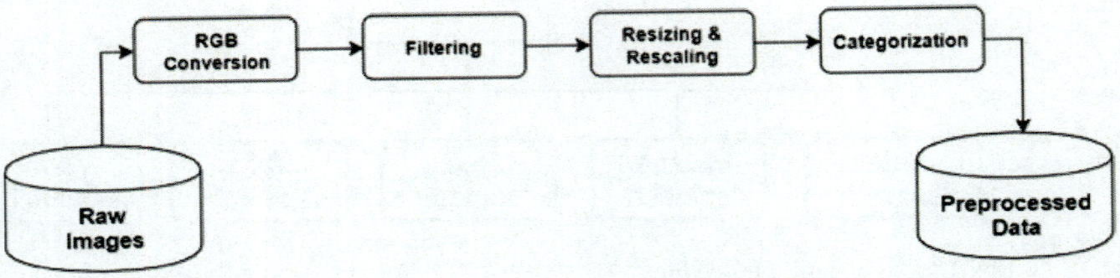

Fig. 3 Pre-processing method

$$x(scaled) = \frac{x - \min(x)}{\max(x) - \min(x)}(nMax - nMin) + \text{nMin} \tag{1}$$

where nMax, nMin is the maximum and minimum value.

4 Experiment and Output

In this section of the article, we will analyze all the results and compare them with each other to see which model is best for our classification and gives the best performance on our dataset.

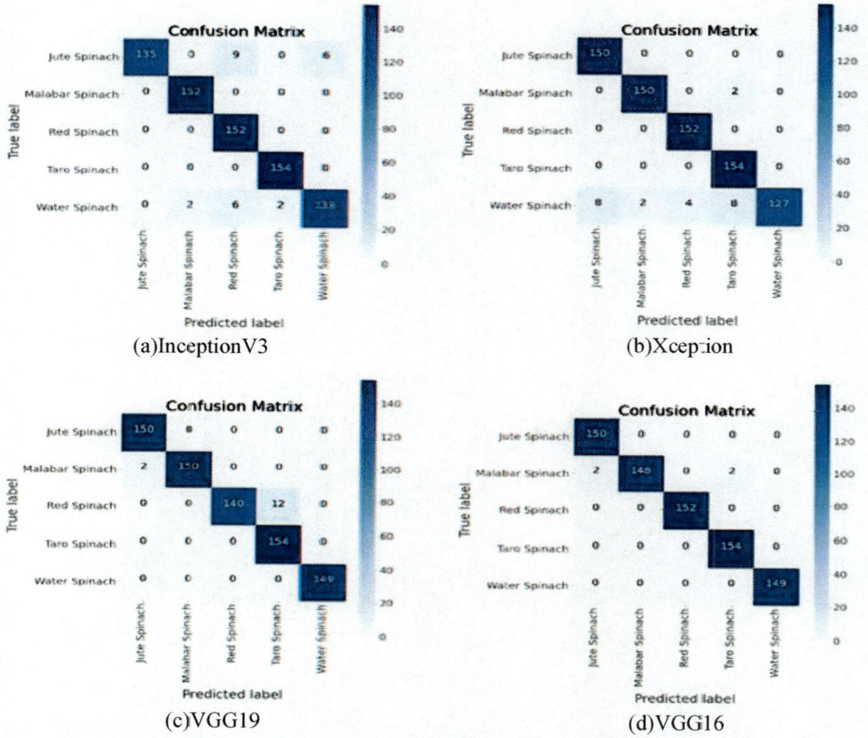

Fig. 4 Confusion matrix for **a** InceptionV3, **b** Xception, **c** VGG19, **d** VGG16

Figure 4 show the confusion matrix for the four CNN models. It is visualized the total correct and incorrect prediction results by the classifier. The predicted label is represented by the *X*-axis and the truth label is represented by *Y*-axis. A confusion matrix contains some number of True Positive (TP), True Negative (TN), False Positive (FP), and False Negative (FN). Suppose here when the classifier predicts it is jute spinach and the actual output is also jute spinach then it is True Positive (TP).

From Table 2 we observed, Malabar and Taro spinach have an accuracy of 99.74. This means that the InceptionV3 model can classify these two spinach better than others. Because of its False Positive Rate 0.33% but False Negative Rate 0%. Contrarily the accuracy of Water spinach has the lowest accuracy of 97.89% because of the total number of False Positive and False Negative 16.

The obtained results by Xception (Table 3) tell us that its accuracy is higher in Malabar and Red spinach classification, which is 99.47%. The number of False Positive and False Negative in both of these is 4. As a result, the sum of False Positive Rate and False Negative Rate is equal. Like InceptionV3, it has the lowest accuracy of the Water spinach class which is 97.10%.

Table 2 Evaluation metric for InceptionV3

Spinach category	True positive (TP)	True negative (TN)	False positive (FP)	False negative (FN)	Precision (%)	F1 (%)	Sensitivity (%)	Specificity (%)	False positive rate (FPR) (%)	False negative rate (FNR) (%)	Accuracy (%)
Jute	135	607	0	15	100	94.74	90.00	100	0	10	98.02
Malabar	152	603	2	0	98.70	99.35	100	99.67	0.33	0	99.74
Red	152	590	15	0	91.02	95.30	100	97.52	2.48	0	98.02
Taro	154	601	2	0	98.72	99.35	100	99.67	0.33	0	99.74
Water	139	602	6	10	95.86	94.56	93.29	99.01	0.99	6.71	97.89

Table 3 Evaluation metric for Xception

Spinach category	True positive (TP)	True negative (TN)	False positive (FP)	False negative (FN)	Precision (%)	F1 (%)	Sensitivity (%)	Specificity (%)	False positive rate (FPR) (%)	False negative rate (FNR) (%)	Accuracy (%)
Jute	150	599	8	0	94.94	97.40	100	98.68	1.32	0	98.94
Malabar	150	603	2	2	98.68	98.68	98.68	99.67	0.33	1.32	99.47
Red	152	601	4	0	97.44	98.70	100	99.34	0.66	0	99.47
Taro	154	593	10	0	93.90	96.86	100	98.34	1.66	0	98.68
Water	127	608	0	22	100	92.03	85.23	100	0	14.77	97.10

Table 4 Evaluation metric for VGG19

Spinach category	True positive (TP)	True negative (TN)	False positive (FP)	False negative (FN)	Precision (%)	F1 (%)	Sensitivity (%)	Specificity (%)	False positive rate (FPR) (%)	False negative rate (FNR) (%)	Accuracy (%)
Jute	150	605	2	0	98.68	99.34	100	99.67	0.33	0	99.74
Malabar	150	605	0	2	100	99.34	98.68	100	0	1.32	99.74
Red	140	605	0	12	100	95.89	92.11	100	0	7.89	98.41
Taro	154	591	12	0	92.77	96.25	100	98.01	1.99	0	98.41
Water	149	608	0	0	100	100	100	100	0	0	100

Concerning the VGG19 results (Table 4), we may notice that it achieved a 100% accuracy for the Water spinach class. Which is more than the rest of the classes. Here Jute spinach, Taro spinach, and Water spinach the Sensitivity of these 3 classes is 100%. With this, Precision and Specificity is 100% for Malabar spinach, Red spinach, and Water spinach. This model has low accuracy for the Red and Taro spinach class, which is 98.41%.

The results are given by VGG16 (Table 5) show that it achieves 100% accuracy for Red and Water spinach. And it has the lowest accuracy for Malabar spinach, which is 99.47%. For the Malabar spinach, Red spinach, and Water spinach class, we observe that it was identified with good specificity and precision which is 100%. And for Jute spinach, Red spinach, Taro spinach, Water spinach class Sensitivity is 100%.

The main performance summary of our experiment has been presented in Table 6. VGG16 has achieved the highest 99.79% accuracy. Out of the total 757 images of the test set, this classifier has been able to classify 753 images correctly. Only 4 images of Malabar spinach class have wrongly predicted, shown in (Fig. 4d). At the same time, it's Precision, F1 score, Sensitivity is 99.47% and Specificity is the highest at 99.87%. This table also tells us that InceptionV3 achieved the lowest accuracy in our experiment, which is 98.68%. In the case of a total of 25 images, the wrong prediction has been made by the InceptionV3 classifier. On the other hand, the Xception model made incorrect predictions for 24 images. As a result, it gave the second-lowest 98.73% accuracy. The difference in accuracy with the InceptionV3 model is only 0.5%. The VGG19 model gives the second-highest 99.26% accuracy. It has classified a total of 743 images correctly and incorrect classification for 14 images, shown in Table 6. By analyzing all these results we can say that the VGG16 model Gives a best performance than the rest in our experiment.

5 Conclusion and Future Work

In this paper, we proposed a local spinach classification method using the CNN model, which can automatically classify the local spinach. Four classifiers proved image processing about similar-looking spinach can be classified and can give a high output through these classifiers. We trained our dataset with several deep learning based convolutional neural networks. Then we used four different models to compare the accuracy between them. All of them gave high accuracy for our dataset. We trained those models with noiseless or few noise images so this model will be unable to give the best performance if the images have noise. As those models gives high performance, we expect that these results will help for further research. Moreover, it will help to develop an image application through the application, general people can classify the edible local spinach.

Table 5 Evaluation metric for VGG16

Spinach category	True positive (TP)	True negative (TN)	False positive (FP)	False negative (FN)	Precision (%)	F1 (%)	Sensitivity (%)	Specificity (%)	False positive rate (FPR) (%)	False negative rate (FNR) (%)	Accuracy (%)
Jute	150	605	2	0	98.68	99.34	100	99.67	0.33	0	99.74
Malabar	148	605	0	4	100	98.67	97.37	100	0	2.63	99.47
Red	152	605	0	0	100	100	100	100	0	0	100
Taro	154	601	2	0	98.72	99.35	100	99.67	0.33	0	99.74
Water	149	608	0	0	100	100	100	100	0	0	100

Table 6 Evaluation metric for four CNN models

Models	True positive (TP)	True negative (TN)	False positive (FP)	False negative (FN)	Precision (%)	F1 (%)	Sensitivity (%)	Specificity (%)	False positive rate (FPR) (%)	False negative rate (FNR) (%)	Accuracy (%)
Inception3	732	3003	25	25	96.70	96.70	96.70	99.17	0.83	3.30	98.68
Xception	733	3004	24	24	96.83	96.83	96.83	99.21	0.79	3.17	98.73
VGG19	743	3014	14	14	98.15	98.15	98.15	99.54	0.46	1.85	99.26
VGG16	753	3024	4	4	99.47	99.47	99.47	99.87	0.13	0.53	99.79

InceptionV3, Xception, VGG19, VGG16

Acknowledgements We gratefully acknowledge for providing GPU support from Computational Intelligence Lab for providing the necessary support. We thank, Dept. of CSE, Daffodil International University. Moreover, thanks to the anonymous reviewers for their valuable comments and feedback.

References

1. Roughani, Mehdi, M.S.: Spinach: an important green leafy vegetable and medicinal herb. In: 2nd International Conference on Medicinal Plants, Organic Farming, Natural and Pharmaceutical Ingredients (2019)
2. Islam, R., Ara, T.: Leafy vegetables in Bangladesh. UBN: 015-A94510112010, Photon ebooks (2015)
3. Wu, S.G., Bao, F.S., Xu, E.Y., Wang, Y., Chang, Y., Xiang, Q.: A leaf recognition algorithm for plant classification using probabilistic neural network. In: 2007 IEEE International Symposium on Signal Processing and Information Technology, vol. 2007, pp. 11–16. Giza (2007)
4. Wang, G., Sun, Y., Wang, J.: Automatic image-based plant disease severity estimation using deep learning. Comput. Intell. Neurosci. **2017**, 1–8 (2017)
5. Sun, Y., Liu, Y., Wang, G., Zhang, H.: Deep learning for plant identification in natural environment. Comput. Intell. Neurosci. **2017**, 1–6 (2017)
6. Zhu, S., Feng, L., Zhang, C., Bao, Y., He, Y.: Identifying freshness of spinach leaves stored at different temperatures using hyperspectral imaging. Foods **8**, 356 (2019)
7. Yalcin, H., Razavi, S.: Plant classification using convolutional neural networks. In: 2016 Fifth International Conference on Agro-Geoinformatics (Agro-Geoinformatics), pp. 1–5. Tianjin (2016)
8. Kaya, A., Keceli, A., Catal, C., Yalic, H., Temucin, H., Tekinerdogan, B.: Analysis of transfer learning for deep neural network based plant classification models. Comput. Electron. Agric. **158**, 20–29 (2019)
9. Priya, C.A.T., Balasaravanan, A., Thanamani, S.: An efficient leaf recognition algorithm for plant classification using support vector machine. In: International Conference on Pattern Recognition, Informatics and Medical Engineering (PRIME-2012), pp. 428–432. Salem, Tamilnadu (2012)
10. Szegedy, C., Liu, W., Jia, Y., Sermanet, P., Reed, S., Anguelov, D.: Going deeper with convolutions. In: IEEE Conference on Computer Vision and Pattern Recognition (CVPR), pp. 1–9. Boston, MA (2015)
11. Szegedy, C., Vanhoucke, V., Ioffe, S., Shlens, J., Wojna, Z.: Rethinking the inception architecture for computer vision. In: 2016 IEEE Conference on Computer Vision and Pattern Recognition (CVPR), pp. 2818–2826. Las Vegas, NV (2016)
12. Chollet, F.: Xception: deep learning with depthwise separable convolutions. In: Proceedings of the IEEE Conference on Computer Vision and Pattern Recognition, pp. 1251–1258 (2017)
13. Simonyan, K., Zisserman, A.: Very deep convolutional networks for large-scale image recognition. ArXiv preprint arXiv:1409.1556 (2014)
14. Dalwinder, S., Birmohan, S.: Investigating the impact of data normalization on classification performance. Appl. Soft Comput. 105524 ISSN: 1568–4946 (2019)

Fusion of Deep Learning Features for Wild Animal Detection

Anamika Dhillon and Gyanendra K. Verma

Abstract A novel method of feature extraction, based on feature fusion, is proposed in this paper. In object classification and detection models, more advantages of transfer learning can be easily extracted from feature fusion. In the proposed scheme, Firstly, features are extracted by using two deep learning models: DenseNet201 and ResNet101 and then a neighborhood component analysis (NCA) based feature selection method is applied to diminish the dimensionality of features. Then the features are fused by applying the concatenation process and finally, SVM is applied to classify the dataset. Experimental results on camera trap wild animal dataset, show that our method can classify the wild animals along with huge variation and extreme occlusion in position and size. Furthermore, we have attained an accuracy of 98.07% which is higher than the algorithm adopting single net and other existing state-of-art techniques.

Keywords Deep learning · CNN models · DenseNet201 · ResNet101 · NCA · SVM · Object detection

1 Introduction

In the area of computer vision, one of the crucial problem is object detection, where the purpose is to detect and segment the target objects in an image. It is widely used for numerous object related applications including pattern recognition, image classification, semantic segmentation, face recognition etc. Utilization of deep learning [1] applications to resolve the problem of object detection [2] become a new area of research. Especially, deep convolutional neural networks (CNNs) [3, 4] are known as the popular techniques to resolve the issues of computer vision, due to its outstanding capability to extract the features automatically. DCNN models extricate the details of features as well as draw the higher level information.

A. Dhillon (✉) · G. K. Verma
NIT, Kurukshetra, India
e-mail: anamika_6170020@nitkkr.ac.in

L. Troiano et al. (eds.), *Advances in Deep Learning, Artificial Intelligence and Robotics*, Lecture Notes in Networks and Systems 249,
https://doi.org/10.1007/978-3-030-85365-5_2

Various feature fusion [5] techniques are proposed in the field of image processing, which have utilized multiple features of an image and then fuse them to achieve better detection performance. All-in-one focused image is achieved by applying the feature fusion technique, where more than one partially focused images are fused together. In this work, we extract the features of the input image by using DenseNet201 [6] and ResNet101 [7]. DenseNet 201 has various compelling advantages: It encourages the reuse of features, it alleviate the problem of vanishing gradient and also enhances the parameter and computational efficiency. However, the speed of training and powerful representational ability is enhanced by utilizing ResNet101. Then we have utilized Neighborhood component analysis (NCA) [8], for visualization as well as reduced-complexity classification. Finally, the features are fused by applying concatenation process and then SVM classifier is applied to classify target image.

The rest of this paper is arranged as follows: Sect. 2 presents the related work. Section 3 provide the details of the propounded framework. In Sect. 4, Experimental Results and analysis are described. In Sect. 5, the conclusion is drawn.

2 Related Work

State-of-the-art techniques for object detection in general classes are generally depend on DCNNs. Animal classification and detection in camera-trap images has recently been considered in [9–11]. In 2019, Han et al. [10] propounded a method based on livestock detection, which utilizes U-net and Google Inception-v4 net and get superior outcome than Faster RCNN and YOLOV3. DenseNet was given by Huang and Weinberger [6], where every layer is densely connected to every other layer in feed forward manner and they make it possible to explore new features. ResNet was propounded by He and Sun [7], it make the utilization of skip connections so that residual functions can be easily learnt. Goldberger et al. [8] propounded a supervised learning method to classify the multi-variate data into several classes. They have learnt a distance metric by finding out the linear transformation of the input data. Laghmati et al. [12] propounded a method for the classification of patients having breast cancer. In this paper, they have used breast cancer Wisconsin dataset and features are selected by NCA. Muhammad Attique Khan et al. [13] have utilized an integrated method of fuzzy C-means and NCA for the recognition of tumor. Three steps are described in their paper: enhancement of tumor visibility, segmentation of tumor and feature reduction through NCA.

3 Proposed Model

Figure 1 shows the mainstream design of the proposed technique. In this article, first of all, the features of input image are learned and extricated by DenseNet 201 and ResNet101 models. Then NCA is applied to the extracted features, so that the best

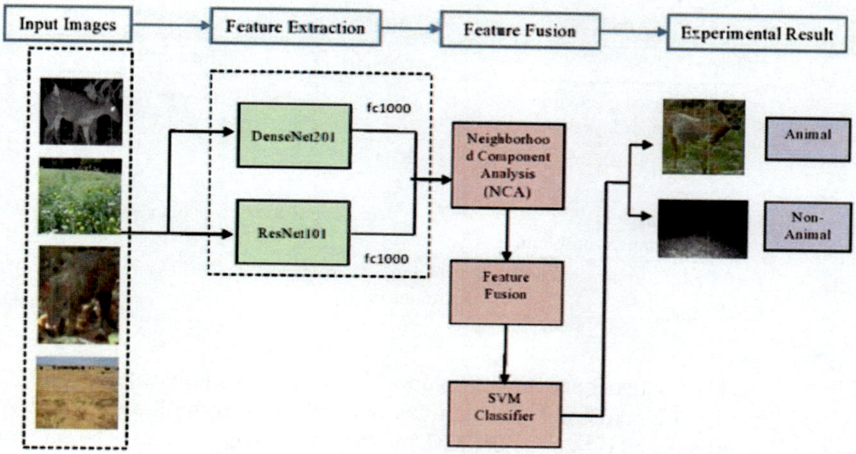

Fig. 1 Flow chart of predictive model

accuracy can be achieved by utilizing the algorithm of stochastic nearest neighbor. NCA authorizes the grouping of data which is outwardly significant, regardless of the huge decrease in dimension. After that concatenation of features is done to fuse them and finally, SVM classifier is conducted for the classification purpose.

3.1 Feature Extraction

We have utilized two pretrained CNN models, DenseNet201 and ResNet101. In DenseNet201, every layer is interconnected directly to the other layer in feedforward way. Whereas ResNet 101 avoid the negative outcomes while increasing the depth. Apparently, DenseNets are quite identical to ResNets and the format of input is non-identical in both networks, which leads to considerably distinct behaviors. The difference in both the networks is presented in Table 1. Therefore, their feature fusion helps the framework in gathering the global as well as local information, to distinguish the animals.

3.2 NCA and Feature Fusion

After feature extraction, neighborhood component analysis (NCA) [14] is applied on the extracted features to select and reduce their dimensionalities. NCA is statistical technique, which creates no supposition about the contour of image classification. It has the major privilege that no information will be lost at the time of feature reduction

Table 1 Comparison between various features for both models

Features	DenseNet201	ResNet101
Format of passing previous layer features to other layers	By concatenating	By summation
Total no of connections in L-layer network	L(L + 1)/2	L
The way to address vanishing-gradient problem	Using dense connections	Using stochastic depth
Performance improvement	Using power of deep architecture	Using power of feature reuse

and also to give the assurance to optical objective function using an optimizer which is based on gradient. NCA used the input set in the form of set of vectors, which is described as: $\{x_1, x_2, \ldots\ldots x_N\}$ where $x_i \in R^m$, and equivalent labels are represented as: $\{y_1, y_2, \ldots\ldots y_N\}$, where $y_i \in L$. A projection matrix A is obtained by NCA, which is of dimension, $p \times m$, where $Q = A^T A$ and it describe Mahalanobis distance as Eq. 1:

$$d(x_i, x_j) = (Ax_i - Ax_j)^T (Ax_i - Ax_j) \tag{1}$$

The probability P_{ij} is assigned for every test point j for assigning the label to its neighbor i. As the distance among i and j increase the probability decreases as shown in Eq. 2:

$$P_{ij} = \frac{\exp(-||\mathrm{Ax_i} - \mathrm{Ax_j}||^2)}{\sum_{k \neq i} \exp(-||\mathrm{Ax_i} - \mathrm{Ax_k}||^2)}, P_{ij} = 0 \tag{2}$$

By utilizing this method, P_i will be calculated easily and i is defined as Eq. 3 and 4:

$$P_i = \sum_{j \in C_i} P_{ij} \tag{3}$$

and

$$C_i = \{j | y_i = y_j\} \tag{4}$$

The final optimization function is described as the correctly classified probability and it is given as: $f(A) = \sum_i P_i$ and finally the gradient of f(A) is described in Eq. 5:

$$\frac{\partial f}{\partial A} = 2A \sum_i \left(P_i \sum_k P_{ik} X_{ik} X_{ik}^T - \sum_{j \in C_i} P_{ij} X_{ik} X_{ij}^T \right) \tag{5}$$

After this dimensionality reduction procedure, features are fused by using concatenation operation, so that they can be learn for the description of their rich internal information. In the CNN architecture fully connected layer act as classifier. In the literature, many articles [15–17] used SVM classifier to classify the data using the features extracted through CNN or deep CNN, due to comparatively better performance than CNN Fully Connected layers for classification. As we have used feature level fusion, we have adopted SVM as classifier for better performance.

4 Experiments and Analysis

4.1 Dataset

In this paper, the images are obtained from the camera trap dataset [18] and captured in both infrared and visual formats and the background in these images is highly cluttered. This dataset contains approximately 5000 images, and the dimension of each image varies in its size. The size of the image varies from 1920 × 1080 to 2048 × 1536, but they are finally resized to 224 × 224. They are categorized into two classes, i.e. positive images and negative images. The positive images contain the wild animals with different background, and in the negative images, the wild animals are not present. From this dataset, 75% of images are captured for training purpose, and the rest 25% are utilized for the testing purpose.

4.2 Results

The software implementation have been performed in MATLAB R2020a platform on Intel core i7 with 8 GB RAM installed. In this phase, we investigate the performance of proposed feature and model fusion techniques for wild animals classification. First of all, training process will occur, where the features are extracted by combining the information from fully connected layers of deep learning models, DenseNet201 and ResNet101. In Fig. 2 illustrates some of the input images in the training phase of our dataset.

Then a nonlinear distance metric learning model NCA is applied on the extracted features and its goal is to maximize the accuracy of regression and classification algorithms. It helps in selecting and reducing the dimensionality of features, which gives us the compact representation of the integrated features. Then in the testing phase, the extracted features are fused by applying the concatenation process and

Fig. 2 Sample of training images. **a** Positive images, **b** negative images

finally, the target images are classified by applying the SVM Model. Figure 3 presents some images during the test phase, whereas Fig. 4 shows some test images with the classification accuracy.

Table 2 illustrates that DenseNet201 achieves an accuracy rate of 94.80% while ResNet101 achieve 93.95% on the wild animal database. Prominently, our method enhances the accuracy to 98.07%, which is 3.27% greater than DenseNet201 and 4.12%higher than that of ResNet101. The proposed methodology certainly can be used in wild photography for animal detection. However, the system must be trained with wild animal database for accurate detection.

The confusion matrix is a graphical display of the classification accuracy on each class of the dataset. The columns represent predicted classes, and the rows represents real classes. For the classification issue of n-class, the confusion matrix is a $n \times n$ matrix. With this matrix the correct recognition rate R_i and the error recognition rate W_i of the data objects can be calculated as:

$$R_i = \frac{n_{ii}}{\sum_{j=1}^{k} n_{ij}}$$

$$W_i = 1 - R_i$$

Here the category set is defined as, $C = C_1, C_2, \ldots C_k$ is given and among them, n_{ii} is the precisely recognized numbers of confusion matrix and n_{ij} denotes that C_i is recognized as C_j by classifier. In Fig. 5 the confusion matrix of the proposed technique is presented.

Comparison of Experimental Results: We compare our model with other frameworks, in terms of Model, Dataset type and accuracy as present in Table 3. We have achieved better results than the current state of the art on the animal dataset with 98.07% accuracy. Basulaim [19] utilizes the NCA method on the Face dataset known

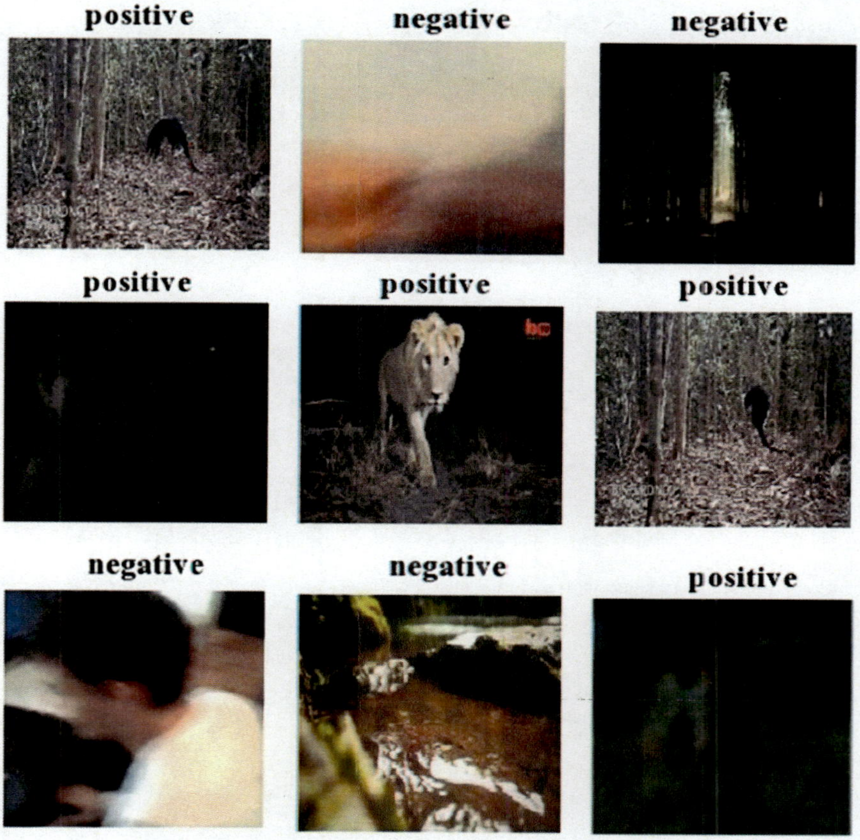

Fig. 3 Images generated during testing phase

as FERET and attains an accuracy of 95.80 whereas Sun et al. [20] achieves an accuracy of 85.33% by applying dimension reduction through PCA and SVM on the Caltech 256 dataset.

5 Conclusion

In recent years feature fusion techniques are employed in computer vision field. In this paper, we present a novel feature fusion technique which is based on two deep learning models, NCA and the SVM classifier. Features are extracted by applying DenseNet201 and ResNet101 models and their dimensionality is reduced by utilizing NCA. Finally the classification process is done by SVM classifier. Experimental results show that our method has attained an accuracy of 98.07% and based on these

Fig. 4 Test images with accuracy rate

Table 2 Comparison between single net and feature fusion of two techniques

Model	Accuracy (in %)
DenseNet201	94.80
ResNet101	93.95
Proposed model	98.07

results we have demonstrated that, our framework is more effectual as compared to other state-of-the-art fusion techniques.

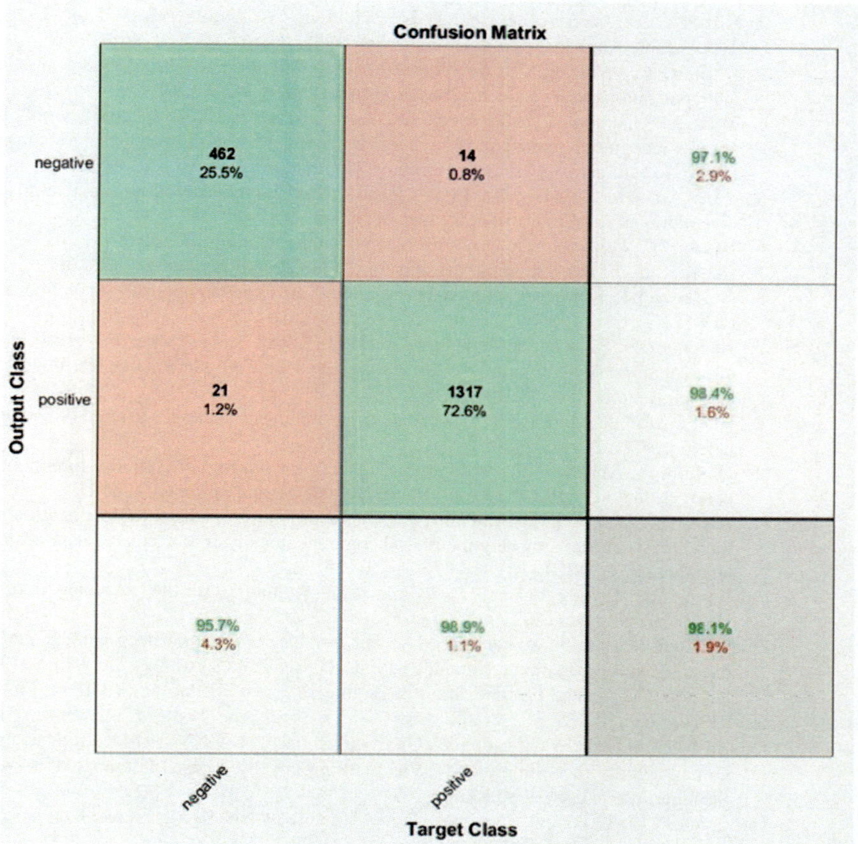

Fig. 5 Confusion matrix for proposed model

Table 3 Comparison with state-of-art methods

Author	Model	Dataset	Accuracy (in %)
Basulaim [19]	NCA	FERET	95.80
Sun et al. [20]	PCA	Caltech 256	85.33
Proposed work	NCA and SVM	Wild animal	98.07

References

1. Liu, L., Ouyang, W., Wang, X., Fieguth, P., Chen, J., Liu, X., Pietikäinen, M.: Deep learning for generic object detection: a survey. Int. J. Comput. Vision **128**(2), 261–318 (2020)

2. Kim, D., Lee, G.: Tell Me What They're Holding: WeaklySupervised Object Detection with Transferable Knowledge from Human-object Interaction (2017)
3. Dhillon, A., Verma, G.K.: Convolutional neural network: a review of models, methodologies and applications to object detection. Prog. Artif. Intell. **9**(2) (2019)
4. Votis, K., Giakoumis, D., Tzovaras, D., Chen, L., Hamzaoui, R.: Feature learning for human activity recognition using convolutional neural networks a case study for inertial measurement unit and audio data. CCF Trans. Pervasive Comput. Interact. **2**(1), 18–32 (2020)
5. Lu, X., Duan, X., Mao, X., Li, Y., Zhang, X.: Feature extraction and fusion using deep convolutional neural networks for face detection. **2017** (2017)
6. Huang, G., Weinberger, K.Q.: Densely connected convolutional networks (2017)
7. He, K., Sun, J.: Deep residual learning for image recognition. pp. 1–9 (2016)
8. Goldberger, J., Roweis, S., Hinton, G., Salakhutdinov, R.: Neighbourhood components analysis (2004)
9. Kellenberger, B., Student Member, Marcos, D., Lobry, S.: Half a percent of labels is enough: efficient animal detection in UAV imagery using deep CNNs and active learning. IEEE Trans. Geosci. Remote Sens. 1–10 (2019)
10. Han, L., Tao, P., Martin, R.R.: Livestock detection in aerial images using a fully convolutional network. **5**(2), 221–228 (2019)
11. Dhillon, A., Verma, G.K.: Wild animal detection from highly cluttered forest images using deep residual networks. Int. Conf. Intell. Human Comput. Interact. (2018)
12. Laghmati, S., Cherradi, B., Tmiri, A., Daanouni, O., Hamida, S.: Classification of patients with breast cancer using neighbourhood component analysis and supervised machine learning techniques. (1) (2020)
13. Rajinikanth, P.V.: Signal and Image Processing Techniques for the Development of Intelligent Healthcare Systems (2021)
14. Siddique MA, Sakib S, Rahman MA.: Performance analysis of deep autoencoder and NCA dimensionality reduction techniques with KNN, ENN and SVM classifiers. In: 2019 2nd International Conference on Innovation in Engineering and Technology (ICIET), pp. 1–6. IEEE (2019)
15. Basly, H., Ouarda, W., Sayadi, F.E., Ouni, B., Alimi, A.M.: CNN-SVM learning approach based human activity recognition CNNSVM learning approach based human activity recognition. Springer International Publishing (2020)
16. Leng, J., Li, T., Bai, G., Dong, Q., Dong, H.: Cube-CNN-SVM: a novel hyperspectral image classification method. 1027–1034 (2016)
17. Jiang, S., Hartley, R., Fernando, B.: Kernel support vector machines and convolutional neural networks. In: 2018 Digital Image Computing: Techniques and Applications (DICTA), pp. 1–7 (2018)
18. Zhang, Z., He, Z., Cao, G., Cao, W.: Scenes using spatiotemporal object region proposals and patch verification. **18**(10), 2079–2092 (2016)
19. Basulaim, K.O.: Performance enhancement for gender recognition using trainable bank of gabor filters and NCA. pp. 0–4 (2019)
20. Sun, Y., Li, L., Zheng, L., Hu, J., Li, W., Jiang, Y., Yan, C.: Image classification base on PCA of multi-view deep representation. J. Vis. Commun. Image Represent. **62**, 253–258 (2019)

Forecast Extreme CPU Usages Under Peak Load Using Envelop EM Semi-supervised Learning

Nitin Khosla and Dharmendra Sharma

Abstract The main objective of an expected maximization (EM) based envelop semi-supervised neural net learning approach in this paper is to develop a model for forecasting peak CPU usages under unpredictable web traffic (load conditions) in a large enterprise applications environment with several hundred live applications. This proposed approach forecasts the likelihood of extreme peak response time because of the stressed CPU due to a burst in incoming web traffic from the live IT applications and then predicts the CPU utilization under extreme load (peak) conditions. The enterprise complex IT infrastructure consists of many applications running simultaneously in real-time. Features are extracted after analyzing the CPU-load patterns of demand which are mainly hidden in the data related to key transactions of the IT applications. This method generates synthetic CPU load profiles by simulating virtual users and execute the key transactions in the test environment. This model is used to predict the excessive peak utilization under peaked CPU conditions. We have used envelope expectation maximization classifier method with forced learning, attempting to extract and analyze the parameters that maximize the likelihood of the model after marginalizing out the unknown labels. This has resulted in mitigating the risks of system failures and enabled us to manage the IT capacity within 3 days from 7 days. This model has helped in IT capacity planning and optimal usages of existing IT infrastructure with minimal risk.

Keywords Semi-supervised learning · Performance and peak load testing · Neural nets · Machine learning

N. Khosla (✉)
Department of Home Affairs, IT Performance Engineering, ETSA Section, Canberra, Australia
e-mail: nitin.khosla@homeaffairs.gov.au

D. Sharma
Computer Science, University of Canberra, Canberra, Australia
e-mail: dharmendra.sharma@canberra.edu.au

L. Troiano et al. (eds.), *Advances in Deep Learning, Artificial Intelligence and Robotics*, Lecture Notes in Networks and Systems 249,
https://doi.org/10.1007/978-3-030-85365-5_3

1 Introduction

Usages of IT applications based upon cloud and data center have grown tremendously in the past decade and our dependence on cloud based solutions is increasing every day in view of new emerging technologies and dynamic user expectations. Frequently the internet traffic or number of hits to an application increases exponentially in a short span of time (internet traffic burst). Due to limited or fixed CPU peak, the peak utilization of the system drastically increases and it slows down the enterprise application system or even breaks down the system at peak load. The IT system crashes at many occasions in an enterprise large environment as the IT infrastructure cannot sustain the excessive load under peak (extreme load) conditions with limited peak. It is observed that many crucial applications halt suddenly which were providing services to the public, *e.g.* custom clearances, air tickets booking, emergency hospital services, and weather forecasting systems in emergencies *e.g.* bush fire, tsunami alerts, etc. Sometimes the IT systems crash randomly due to unpredictable high load or high volume of internet traffic and with limited peak availability, this results in degradation of efficiency and adversely impacts the productivity. Many times, the system alerts are not observed and practically it is very challenging to for large enterprise organizations to take any remedial actions *e.g.* load balancing, etc. IT system becomes irresponsive and unable to process large number of transactions requests because of extremely high transaction rate which peaks randomly. Managing the keys applications running 24/7 at a high efficiency level is a constant challenge between functionality and resource management [1] in a big enterprise system. Sometimes very little or no peak is available for the critical applications to process transactions in an enterprise set-up and it can lead to a system crash. These circumstances are even more complex when transactions are generated due to wide area distributed CPUs where the network latency and peak web traffic are key factors impacting the performance of applications.

2 Research Objective

Primary objective of this research paper is to design and implement a labelled based envelop expected—maximization semi-supervised neural net approach to predict the usages of CPU (peak) utilization under unpredictable high volume of internet traffic.

Work-load profiles and patterns were studied to identify CPU (peak) utilization patterns at different time periods during last one year. These patterns were used to develop test scenarios and tests were designed in the test environment for the validation purpose. We also collected the test data, analysed and extracted load patterns from the logs of raw transactions. Extreme CPU utilization was traced using data mining and this enabled us to identify the bottlenecks and helped us to estimate a projected load to carry out experiments in test environment. We used this information to forecast expected CPU (peak) utilization for the real world (production) environment. We developed a predictive model to forecast the CPU (peak) performance under extreme peak conditions (stress) in a big enterprise applications set-up.

3 Enterprise Environment and Issues

Large public service departments and big corporates comprises of different types of system architectures (latest and legacy) *e.g.* smart mobile applications, cloud computing, etc. Our experimental data was collected from a big e and complex environment consists of more than 350 servers. Most of these servers were distributed across multiple sites (countries). Our validation experiments were performed in a test environment (pre-production environment), which was a subset of the whole enterprise set-up containing all of the applications with the most latest builds but with limited data set. This test environment was also used representing a system simulating all the applications distributed in more than 52 overseas posts across the world.

3.1 Performance Issues in Current Practices

Computer applications are developed based upon user demands and expectations and the expectations are primarily dynamic in nature. Users ever changing demands pose many critical limitations in evaluating the performance of IT applications in the current practices such as,

- Risk Based Approach: Performance testing is mainly done on key (high priority) transactions or on the high-risk use-cases only because testing every scenario or combinations is extremely time consuming and costly. So hidden defects cannot be found.
- Not Robust: Lack of a practical and robust approach which could provide useful results in short time frames. It is mainly due to the volatile and unpredictable web traffic.
- Not Reliable: System's response prediction *e.g.* CPU response time, performance, etc., under high volume of web traffic is not reliable.

4 Workload Patterns and Features Extraction

Profile points were created and configured at different layers in the IT infrastructure and these profile points stores data continuously on pre-defined time intervals. We collected the transactional data from data logs/files which were generated at fixed

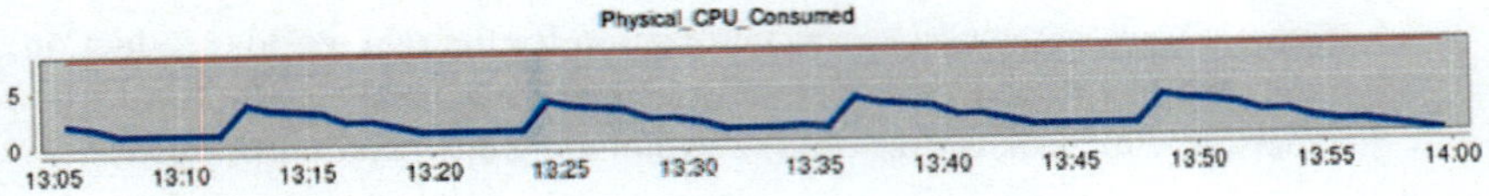

Fig. 1 Percentage CPU usages on a server

periodic intervals. Different transactional data was captured e.g. TRT (transaction response times), percentage peak utilization, memory usages, etc.

Performance and peak experiments for validation were done in IT test environment which represent production like environment called *"pre-production environment"*. The validation testing environment configured in such a way that it emulates the real-world scenarios or system behaviour. Profile points were used to monitor the transactions, collect responses time data and stores the key transactional parameters during the server-response path both (server to client and client to server). These CPU utilization were then represented by a time series consisting of a pattern and/or a cyclical component.

Figure 1 shows a CPU work-load patterns which follows a cyclic pattern. We captured the data during the two peak intervals where we could observe a typical pattern. We have also collected some data related to hard disk usages, database hits, memory usages, etc. during the peak CPU utilization and did some analytics to find insights from these patterns for predictive modelling. Hierarchical dependence and the impact of secondary transactions are out-of-scope and will be investigated as an extension to current work.

5 Envelop EM Semi Supervised Learning

We used a envelop based EM semi-supervised learning method to train our model. In this approach we used labelled data along with some amount of unlabelled data [2]. Some of the key advantages associated with this research work are,

- It is a reliable and scalable predictive approach.
- It has generated a forecasting model simulating analogies of CPU load patterns based upon captured data from different data profile points.
- Optimising efficiency in terms of response time and accuracy by timely predicting trends which could avoid system failures.

Assumptions: In order to our EM semi-supervised learning approach to work, we have assumed some assumptions e.g. if two distinct points d_1, d_2 are close, then there might be respective outputs y_1, y_2. Without these assumptions, it would be challenging to develop a practical model with training data to forecast a set of infinitely number of test-cases which are mainly unseen or unpredictable [3, 4].

We used some useful parameters such as—effort, time, tools limitations and resource utilization. In view of the potential implementation of the outcome of this research work, the envelop EM semi-supervised learning along with forced-training method [1, 5] has provided some useful outcomes because it is based upon,

- Assumptions of forced training can reduce the training time.
- Training of data set with both labelled and unlabelled data.

Let the feature vector x (1 dimensional) is represented by $d \times 1$ representing the utilization pattern of CPU usages in the real-world environment. Let X be the $J \times d$ feature matrix of the labelled samples and X_u be the matrix $U \times d$ of unlabelled samples. Let y be the $L \times 1$ vector containing the labels represented (encoded) between $\{0, 1\}$ showing the features of % CPU usages and let w denotes the weight vector of our classifier.

$$L_s(w) = \sum_{i=1}^{J} \left(x_i^T w - y_i\right)^2 + \mu w^2 \tag{1}$$

where minimizer of the above objective function is given by,

$$w = \left(X^T X + \mu I\right)^{-1} X^T y \tag{2}$$

A new object is labelled with a threshold of ½,

$$c_w(w) = \begin{cases} 1, & x^T w > \frac{1}{2} \\ 0, & otherwise \end{cases} \tag{3}$$

A variable u is introduced in the objective function (Eq. 1) to include the unlabelled objects. The modified objective function becomes,

$$L_u(w, u) = X_c w - \begin{bmatrix} y \\ u \end{bmatrix}^2 + \mu w^2 \tag{4}$$

where X_c is the concatenation of X and X_u.

After taking the gradient of the function (Eq. 4) and minimizing labelling, each label representing the percentage CPU peak utilization can be projected onto $\{0, 1\}$.

6 Experiment Setup and Validation Methodology

Our CPU patterns represented the real-world type scenarios by simulating different transactions at different time intervals. The test environment has a sub-set of data which represents large data associated with the integrated applications in real word environment. About 139 live and fully functional applications were installed in the test environment representing the real applications environment. This process included,

- Data collection and feature extraction of the of workload demand patterns during last one year at regular short intervals.
- Generate synthetic workloads patterns in our test environment.
- Run stress and peak tests in the test environment with the help of large number of virtual users using the applications just as a real-world scenario.
- Validate the results obtained by extracting data from different profile points configured at application threads, nodes and CPU layers in the integrated environment.
- Train the envelop EM model using semi-supervised learning approach (deep learning paradigm Expected Maximization) [6].
- Forecast the likelihood of CPU extreme usages due to burst in the web traffic [7, 8].

We implemented a virtual load experimental setup in our test environment for performing load experiments and validation.

(i) VU (Virtual User) Generator: these VR users were used to simulate key end-user business transactions.
(ii) Controller System: To manage, control and monitor the execution of load tests with specified ramp up and ramp down strategies.
(iii) Load Generators: These were set up on servers to generate virtual load and to simulate the CPU workload load patterns while using large number of virtual users. These users simultaneously generated web-traffic bursts.

Figure 2 shows a CPU load profile with a typical group of virtual users (under peak conditions) with different slopes of ramping up and ramping down times of VU users. The above set up was used for validation of our results in the test environment.

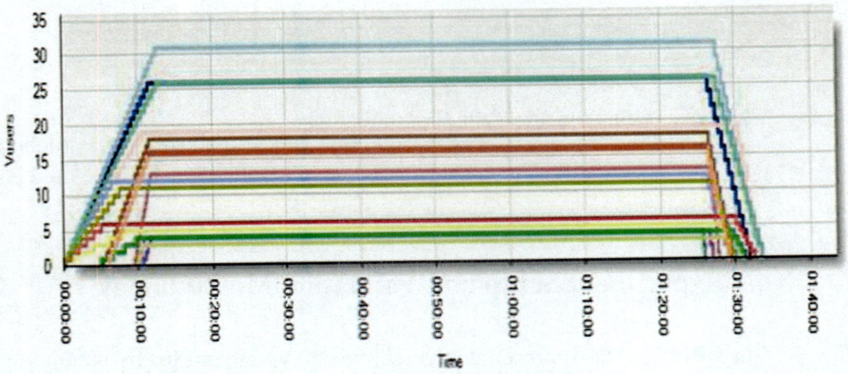

Fig. 2 Load profile (virtual users with different ramp up and down)

7 Forecast CPU Usages

We have worked out the aggregate demand difference of each occurrence of the pattern from the original workload and compared. We have used the modified exponential smoothing (ETS) algorithm with ETS point predicts are equal to the medians of the predict distributions. This approach has enabled us to analyse the trend in CPU usages (Fig. 4).

Table 1 shows results when we executed the tests with different learning algorithms. It was observed that envelop EM semi-supervised learning algorithm (NDSM) has minimum percentage least square loss.

Figure 3 shows the results of a labelled envelop EM semi-supervised neural CPU model (EM), which is used to predict the percentage CPU usages under burst of web traffic [3, 9].

Our forecasting model and new approach is now a part of our monitoring and resource management process and used to evaluate work load demand patterns in

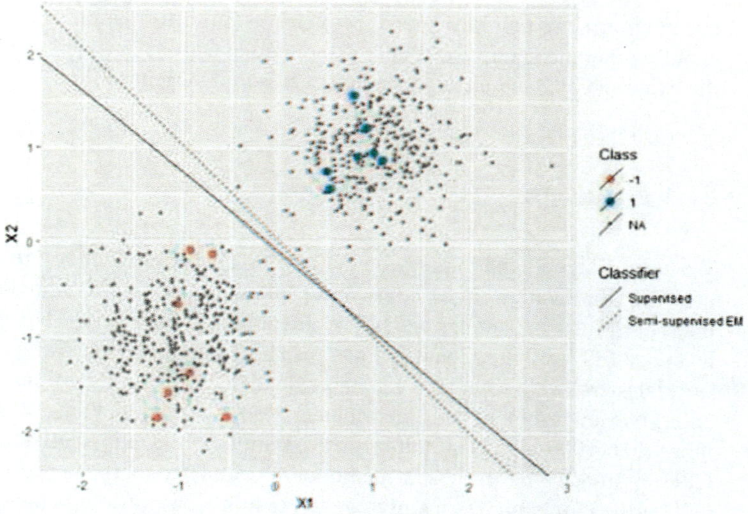

Fig. 3 Envelop EM semi-supervised versus supervised learning

Table 1 Comparison of different machine learning algorithms

Learning algorithms	Weight decay	% Mean square loss
Supervised least square classifier	0.1	15.415
Supervised EM nearest mean classifier	0.1	13.0554
Semi-supervised EM learning classifier	0.1	11.5892
Semi supervised envelop classifier (NDSM)	0.1	11.0056

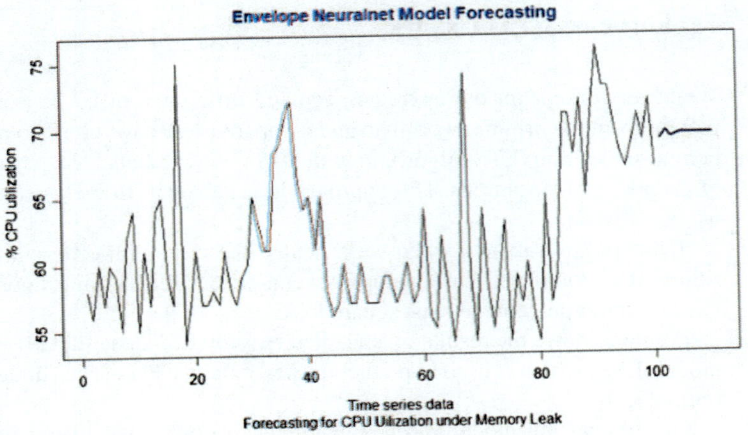

Fig. 4 Percentage CPU utilization forecast under peak load

our environment. Our new model provides useful information to system architects enabling them to set up alerts and take remedial actions in time *e.g.* re-allocation of the existing CPU and memory resources to avoid system failures.

8 Conclusion

We have designed and implemented a new practical approach to predict percentage peak CPU utilization under stress and peak conditions with unpredictable burst in internet traffic in a complex and highly integrated environment. The test environment had over 139 fully functional and live web based applications. Several thousand VU (virtual users) were simulated generating load under stress conditions. The integrated enterprise environment had a distributed system with more than 350 servers serving more than 1000 clients simultaneously. Using the envelop semi-supervised neural CPU approach, the proposed methodology predicts sharp increase in percentage CPU utilization which is mainly due to burst in web-traffic in a complex enterprise IT infrastructure. Data analytics based upon our experiments enabled the system architects and IT system capacity planners to distribute the load appropriately and perform load balancing in time. It is observed for last 6 months that this model has reduced the risk of potential systems failure and improved the system performance and effectiveness.

This new practical approach helped to implement mitigation strategies which were implemented in few hours (3–4 h) compared to about one week in previous practice. Validation of our results were done in an integrated test environment and the model was used to generate alerts as soon as the CPU utilization of the combined server's crosses 75% critical limit. This approach has proven beneficial for our department in managing IT resources, planning IT capacity, optimizing IT resources in the complex enterprise IT environment.

9 Future Work

As further work, we have commenced working on a hierarchical envelop EM semi-supervised learning model to forecast workload patterns. We will consider other parameters *e.g.* hard disk failures, database reads/writes and complex transactions and design an accelerated envelop EM semi-supervised learning approach for predictive modelling.

References

1. John, K., Michael, S.: Preventing memory leaks in windows applications. In: Microsoft Windows Documents (2018)
2. Iqbal, M.F., Zahid, M.Z., Habib, D., John, K.: Efficient prediction of CPU traffic for real time applications. J. Comput. CPUs Commun. **2019** (2019)
3. Zhao, H., Ansari, N.: Wavelet transform based CPU traffic prediction: a fast online approach. J. Comput. Inf. Technol. **20**(1) (2012)
4. Nie, L., Jiang, D., Yu, S., Song, H.: CPU traffic prediction based on deep belief CPU in wireless mesh backbone CPUs. In: IEEE Wireless Communication and CPUing Conference, USA (2017)
5. Yu, C., Wang, D., Yang, T.: Adaptive shaping reinforcement learning agents vis human reward. In: PRICAI Proceedings Part-1, Springer (2018)
6. Kingma, D.P., Rezende, D.J., Mohamad, S., Welling, M.: Semi-supervised learning with deep generative models. In: Proceedings of Neural Information Processing Systems (NIPS), Cornell University, USA (2014)
7. Wang, X., Zhang, M., Ren, F.: DeepRSD: a deep regression method for sequential data. In: PRICAI Proceedings Part-1, Springer (2018)
8. Gmach, D., Rolia, J., Cherkasova, L., Kemper, A.: Workload analysis and demand prediction of enterprise data center applications. In: IEEE 10th International Symposium on Workload Characterization, Boston, USA (2007)
9. Diederik, K., Danilo, R., Shakir, M., Welling, M.: Semi-supervised learning with deep generative models. In: Proceedings of Neural Information Processing Systems (NIPS) (2014)
10. Oliver, A., Odena, A., Raffel, C., Cubuk, E.D. et al: Realistic evaluation of semi-supervised learning algorithms. In: 6th International Conference on Learning Representations, ICLR, Vancouver, BC, Canada (2018)
11. Liu, Y., Krichhoff, K.: Graph based semi-supervised learning for phone and segment classification. Interspeech, France (2013)
12. Rezende, D.J., Mohamed, S., Wierstra, D.: Stochastic back-propagation and approximate inference in deep generative models. In: Proceedings of the 31st International Conference on Machine Learning, Beijing, China (2014)
13. Sankar, A., Zhang, X., Chang, K.C.-C.: Meta-GNN: metagraph neural CPU for semi-supervised learning in attributed heterogenous information CPUs. In: Proceeding of the IEEE 2019/ACM International Conference on Advances in Social CPU Analysis and Mining ASONAM (2019)

14. Miyato, T., Maeda, S., Koyama, M., Ishii, S.: Virtual adversarial training: a regularization method for supervised and semi-supervised learning. IEEE Trans. Pattern Anal. Mach. Intell. **41**(8) (2019)
15. Bastidas, G., Abdai, S., Shah, N., Papalexakis, E.E.: Semi-supervised content-based detection of misinformation via tensor embeddings. IEEE/ACM International Conference on Advances in Social Networks Analysis and Mining (ASONAM) (2018)

Classifying Case Facts and Predicting Legal Decisions of the Indian Central Information Commission: a Natural Language Processing Approach

Arvind Kalia, Naveen Kumar, and Nischay Namdev

Abstract One of the prime challenges the legal organisations around the world facing is the challenge of reducing the pendency of lodged cases. In India, the problem of the pending legal cases is severe and has led to delayed justice. The study addresses the issue by proposing a framework which can help predict the case outcome using case facts. The case data of the Indian Central Information Commission has been used for predicting the outcome of the case in terms of violation or non-violation of the Parliament of India RTI Act. The case facts have been extracted from the scrapped case documents of the Indian Central Information Commission. The study defines the prediction of case outcome as a binary classification problem. Using three feature representation methods (Bag of Words, Word2vec and Sentence-BERT) and four learning models (Support Vector Machine, Naive Bayes, K-Nearest Neighbour and Decision Tree) the performance has been compared and analysed utilising accuracy metric. The study will also put some light in the performance of the feature representation methods described above when used with the different machine learning models. It is one of the first studies performed in the field of the prediction of the outcome of the legal cases in the Indian legal context. Therefore, it has been anticipated that the study shall fill some gap between two fields of Machine Learning and Law.

Keywords Artificial intelligence · Machine learning · Natural Language Processing · Law

A. Kalia · N. Kumar · N. Namdev (✉)
Mandi, HP, India

L. Troiano et al. (eds.), *Advances in Deep Learning, Artificial Intelligence and Robotics*, Lecture Notes in Networks and Systems 249,
https://doi.org/10.1007/978-3-030-85365-5_4

1 Introduction

Legal organisations throughout the world produce seer amounts of data in the form of case documents. As the cheap data storage and networking technologies grow, the amount of legal data produced by such organisations grows. The average number of cases instituted from 2009 to 2018 in the Indian Supreme Court was around 75,000 cases per year [1]. Such a vast quantity of legal cases result in numerous pending cases in the courts leading to delayed justice. But large amounts of data produced by courts can be used to develop machine learning models, which can help solve the problem of pendency of cases in the courts by predicting case outcomes. The study envisages that a judge can use such models as a supplementary tool to fasten the process of adjudication and the lawyers to assist their clients. The study presents an approach which uses Natural Language Processing and Machine Learning methods to extract and classify case facts to predict the outcome of the Indian Central Information Commission cases. The Indian Central Information Commission (ICIC or CIC) is a statutory body set up under the Parliament of India Right to Information Act in 2005 [2]. The Commission has powers to adjudicate an appeal filed under section 19 (3) of the RTI Act and order for providing the requested information when the same has not provided at the level of Central Public Information Officer (CPIO) or the First Appellate Authority [3]. The goal of the study is to predict if there is any violation of the RTI Act or not using case facts extracted from the CIC case documents. The study claims that published case facts labeled as violated or not-violated according to the RTI Act can help predict the outcome of the future case in terms of the violation of the RTI Act. It is very similar to the principle of legal Precedent. Common-law legal systems place great value on deciding cases according to consistent principled rules so that similar facts will yield similar and predictable outcomes. Therefore, Judges have to follow precedents. This principle is also called *Stare Decisis* [4].

The problem of predicting the CIC cases has been considered as the binary classification problem. The case facts extracted from the case documents scraped from the CIC database have been used to build the learning models. Three different textual feature representation methods utilized were:

- Sentence-BERT
- Word2vec
- Bag of Words N-grams with TF-IDF values.

Features generated using the above methods were then used to train Naive Bayes, Decision Tree classifier, K-Nearest Neighbour and three different kernels of SVM (Linear, Radial Basis Function and Polynomial). Comparison of the performance using accuracy metric of the different models shows that Linear SVM with n-gram features still outperforms the state-of-the-art Sentence-BERT and Word2vec by large numbers.

2 Related Work

One of the earliest works done in the field of the prediction of the outcome of legal cases uses simple statistical forecasting methods. Martin et al. used the statistical forecasting model based on information derived from past U.S. Supreme court decisions to construct a classification tree and compared it with the cases classified by legal experts. It was observed that the statistical method outperformed legal experts by 16.8% [5]. Brüninghaus and Ashley presented the method which used integration of SMILE (Smart Index Learner) + IBP (Issue-Based Prediction) to predict the case outcomes [6].

Most influential of the works done in this field was by Aletras et al. Using Bag of words model and topics in the SVM model, the framework could predict the outcomes of the European Court of Human rights cases with an overall accuracy of 79%. It was observed that the facts section of the document helps majorly in predicting the outcome compared to the other sections [7]. Sulea et al. were the first to apply text classification method to legal texts from the French supreme court and to carry out the temporal text classification on a legal text for the first time. The study used BOW model and SVM classifier to predict the outcomes of the cases. Three main goals of the study were to predict the law area of the case, the court ruling based on the case description and estimation of the data of the ruling. The study could achieve a 90.2% F1 score on law area prediction [8]. Virtucio et al. proposed a model to predict the Philippine Supreme Court decisions highly influenced by [7]. Extracted n-grams and topics were used to train SVM to create a learning model. SVM predicted 45% of decisions on n-grams whereas 55% for the topics. But Random forest classifier gave the highest accuracy of 59% for decision prediction on n-grams [9].

However, the latest trends suggest that deep-learning is playing a major role in predicting legal cases [10]. Xiao et al. introduced a large scale legal dataset for judgement prediction. Dataset was part of the Chinese AI and Law challenge dataset (CAIL2018). Dataset contained over 2.6 million criminal cases published by the supreme court of the people's republic of china. Li et al. applied a CNN (Convolutional Neural Network) for legal text classification used for prediction of the outcome of cases. The Study used Word2vec embedding as input to train CNN. It was concluded that the model was efficient than the traditional neural network models [11]. Yan et al. proposed a model for law article prediction based on deep learning. It was developed using TEXT-CNN model which uses a neural network based model [12]. The study utilised processed fact descriptions of the cases as input and law articles as labels. The CAIL data set was used for model creation. Comparison of the Text-CNN performance with the traditional models showed that Text-CNN was better than all the traditional models with an F1 score of 86.27% [13].

Work in the field of legal case outcome prediction can help solve the greater challenge of pendency of cases in courts throughout the world. It will result in faster justice and efficient legal systems. Judges can use such systems as an assistive tool to deliver justice efficiently and swiftly. This research is of one the first studies to devise a framework for the prediction of legal case outcome of any Indian judicial organisation, in this case, it is the Indian Central Information Commission (ICIC or CIC). It is believed that the study will act as a catalyst for various corporations and researchers trying to solve the real-world problems related to the legal domain.

3 Methodology

The methodology used in the study has been divided into four sections as depicted in Fig. 1. Data Gathering section gives details about the methods used to get the data from the database of the CIC. Data Preparation and Labeling discusses the methods used for data pre-processing and labeling. Finally, the last two sections describe the feature extraction methods and classification models used in the study.

3.1 Data Gathering

The study uses the data gathered from the database of the Indian Central Information Commission (ICIC or CIC) [2]. The Indian Central Information Commission was constituted with effect from 12-10-2005 under the Right to Information Act, 2005. The jurisdiction of the Commission extends overall Indian Central Public Authorities. CIC entertains the second appeal cases related to Right to Information (RTI) Act 2005. The RTI act is an act of the Parliament of India. The RTI Act enables citizen's right to information from public institutions. Through the RTI Act, any Indian citizen can ask for information from most of the public institutions other than a few exceptions related to military, nuclear etc. [14].

The case decisions of CIC extracted from their database through the web were utilised in the study [2]. The web interface of CIC provides the functionality of sorting the case decisions in four ways as following:

- Type—*single judge decision, full bench decision, Dispatch decision*
- If the Single judge Decision is selected, there is an option to choose the judge responsible for that decision
- Type of file—*Appeal or complaint*
- Decision type—*Interim, Main, Corrigendum, Compliance order*.

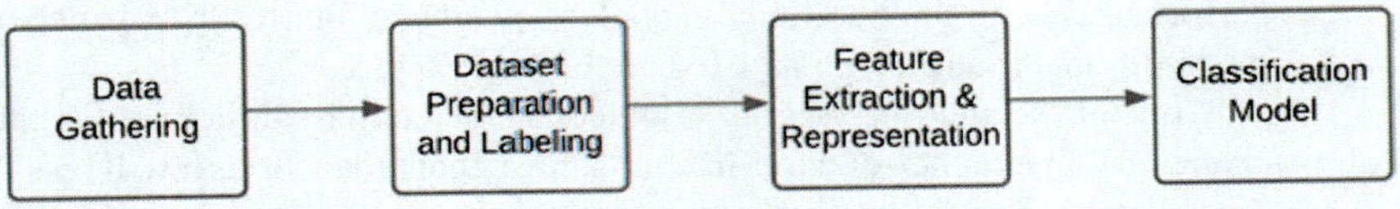

Fig. 1 Steps involved in the methodology

The options used were:

- Type of Decision—*single judge decision*
- Type of File—*Appeal*
- Decision type—*Main*.

Main decisions mean that they are the final decisions of the CIC and there can be no further appeal. Documents gathered from the database were in the PDF's form (Portable document format). About 1300 case documents of eleven judges of the CIC were collected. It could have been that the gathered case documents were of the single judge only but this would have resulted in an inherent bias towards that particular judge resulting in lesser efficiency of the classification model. Therefore, it was reasonable to consider all the judges. The decision documents were stored in directories named after the judge to which the decision belonged.

3.2 Dataset Preparation and Labeling

Now, the case documents were stored in directories in the PDF form sorted according to the judges. But the larger challenge lied ahead. The challenge of labeling the PDF files as violation or non-violation of RTI act. For this task, The knowledge of the domain expert has been picked (in this case, a lawyer). Domain expert read all the 1300 files going through different directories of judges. They then sorted the decisions based on their domain expertise. The judge directories were further extended with two more directories (violation and non-violation) added to them. The violation directory contained decisions of the judge which were the violation of the RTI act. The non-violation directory contained the decisions of the judge which were the non-violation of the RTI act. After the above step, there were only 1108 decision documents left in the pdf format. Next step involved extracting the case facts from remaining 1108 case documents. It was achieved by manually going through each case document and extracting case facts from it. One important thing to note was that some case documents contained the decision from the lower court. In the study, it was considered that the previous court decisions constitute the facts of the case. And according to the Stare Decisis principle, the final decisions should be derived from the case facts and the previous hearing/decisions of the lower court. Finally, all the documents were transformed into text format files. The text was then loaded into a Pandas data-frame column with the second column as the labels [15]. The label zero (0) represented non-violation and one (1) violation. After removing the blank or NaN values in the dataset there were only 1105 instances left in the dataset. The total number of violation text remained was 561 and non-violation was 544. It is worth noting that there was negligible class imbalance in the dataset [16]. The difference of instance between the two classes were only 16.

3.3 Feature Extraction and Representation

The study uses three different feature vector representation methods. Feature vector representation refers to transforming textual data into a numerical vector so that a machine learning model can understand it. These three methods have been compared and analysed by the performance when applied to a machine learning classifier.

Bag of Words Model with TF-IDF Values Bag of words (BOW) model has been used to get numerical features from the textual data. BOW is a model which describes the occurrences of words in corpora. BOW model represents text as a multi-set of its words (either uni-grams or N-grams). The BOW model discards grammar and word order of the text. These numerical features can be used as an input to a machine learning model [17]. The study uses the Term Frequency-Inverse Document Frequency (TF-IDF) approach to represent the text instead of a simple word count. Term Frequency-Inverse Document Frequency (TF-IDF) is a measure of how important a word is to a document in corpora [18]. TF-IDF has shown better results in machine learning models when compared to the simple count frequency used in the BOW model. For the study, only n = 1 and n = 2, n-grams were used. Textual data was tokenised into words. The stop words and numerical characters from the corpora were removed to reduce noise. TF-IDF values can be calculated as:

$$TF = \frac{\text{Number of times term t appears in a document}}{\text{Number of terms in the document}} \tag{1}$$

$$IDF = \log\left(\frac{N}{n}\right) \tag{2}$$

where N is the number of documents and n is the number of documents a term t has appeared in.

Word2vec Word2vec is a neural-network-based pre-trained feature extraction method. It uses a neural network to learn word associations in a corpus of text. Word2vec represent each word as a numerical vector of length one hundred. Mathematical similarity functions (for. e.g. cosine similarity function) were used to create a vector for each word. Therefore, the vector represents the similarity between words in the text. Word2vec either uses skip-gram or continuous skip-gram model [19]. The study uses the average of each word vector which is of length 100 in each of the 1105 instances of the dataset and creates a single vector for each instance to feed it to the machine learning model. The shape of the final matrix was (1105, 100).

Sentence-BERT Sentence-BERT (Bidirectional Encoder Representations from Transformers) is a state-of-the-art feature extraction approach in Natural Language processing. Instead of using words, Sentence-BERT uses sentences to extract features from the corpora of text. For each sentence it produces a vector of length 768 [20]. The study uses the average of the multiple sentence vectors generated from the case text to form a single vector of length 768. The shape of the matrix used as a input to the machine learning model was (1105, 768).

3.4 Classification Model

For the study, six learning models were created:

- Three kernels of SVM (Linear, Polynomial and RBF kernel) [21].
- Naive Bayes [22].
- K-Nearest neighbour [23].
- Decision tree classifier [24].

The comparison and measurement of the performance were done using an accuracy metric. The data has been split in an 80:20 manner. 80% of data was used for training the models and remaining 20% for testing. Later, three feature representation methods described in the previous section were applied in each of the learning models for comparative analysis. For training the model, the non-violation text was labeled as '0' and violation text as '1'.

4 Results and Discussions

This section presents the results and discussions. It also describes why the accuracy metric was chosen over the F1 score for the performance analysis.

4.1 Comparative Analysis of Performance

The performance of each of the six models with different feature representation methods has been analysed using accuracy metric. The reason for the usage of the accuracy as a performance metric was that accuracy performs better when there is a little or no class imbalance and true positives and true negatives are more important. F1 score could have been utilized for measuring the performance of the models, but the F1 score performs better when there is a huge class imbalance in the data and false negatives and false positives are more important. The accuracy can be easily calculated using confusion matrix [25]. The accuracy of a model can be calculated as:

$$accuracy = \left(\frac{TP + TN}{TP + TN + FP + FN}\right) \quad (3)$$

Table 1 Accuracy of learning models

Classifier	Kernel	BOW	Word2vec	Sentence-BERT
SVM	Linear	70%	53.4%	66.0%
	Polynomial	66.0% ($d = 2$)	55.5% ($d = 3$)	65.3% ($d = 6$)
	RBF	64.6%	54.8%	62.0%
Naive Bayes	–	64.6%	47.2%	59.2%
K-NN	–	67.5% ($K = 1$)	58.1% ($k = 5$)	63.8% ($K = 1$)
Decision Tree	–	64.9%	50.1%	53.4%

where TP is the number of true positives, TN is the number of true negatives, FP is the number of false positives and FN is the number of false negatives. In simple terms:

$$accuracy = \frac{\text{number of correct predictions}}{\text{total number of predictions}} \tag{4}$$

The accuracy of each of the model with different feature representing methods is depicted in Table 1.

From Table 1, it was observed that the linear kernel of SVM with BOW outperformed all the other learning models and feature representation methods. The three feature representation method columns (BOW, Word2vec, Sentence-BERT) in the above table, depicted that linear kernel of SVM outperforms others in BOW and Sentence-BERT column whereas K-NN with k = 5 outperforms others in Word2vec column. While Word2vec recorded the lowest performance among all feature representation methods. Delving deeper into the best performing model and analysing the predictive accuracy of each class. It was observed that Violation cases have better predictive accuracy than non-violation cases. For violation cases, it achieved the accuracy of 76.8% and for non-violation cases, the accuracy was 61%. It was also analysed the features with the most positive and most negative weight in the learning model. Table 2 shows the top 10 most positive and negative weight features.

4.2 Discussions

The results suggested that linear kennel of SVM with BOW still outperforms the state-of-the-art Sentence-BERT. Aletras et al. used the linear kernel of SVM with BOW features and was able to achieve the accuracy of 79% [7]. It is also clear from the results that the accuracy of the violation cases is better than the non-violation cases by huge numbers in the best performing model. So, violation cases were simple to predict than the non-violation cases. The top 15 predictive features also validate the results in somewhat manner. The top 15 negative features include words like dissatisfied, required, ordered, etc. The top 15 positive features include words like

Table 2 The top 15 positive and negative weight features

Top 15 positive predictive features	Top 15 negative predictive features
Information	Required
Ranchi	Ordered
Action taken	Appropriate reply
Information appellant	Cic
Loan	Submitted appropriate
Loans	School
RTI application	Manager
Provide	Pension
Eco bank	Provided
Direct respondent	Appellant present
Appellant submitted	Provided appellant
Regarding	Vide
Submits	Dissatisfied
Providing information	Reply provided
Police	Appellant remained

information, the action is taken, provide, direct respondent, etc. These words suggest one or the other class to which these features belong. It has been also observed that removing nouns from the case facts can help improve the predictive accuracy of each class. Nouns like names of places and cities were irrelevant in deciding the case decision but they acquire the place in the top 15 predictive features. Removing nouns can give clarity and improve the accuracy of the classification model. Virtucio et al. also suggested that by preparing the dataset in a fully automated way could help improve the performance of the model, therefore the way the study handles the dataset preparation limits the scalability as well as the data that can be extracted from the case document [9].

5 Conclusion

This was one of the first studies on prediction of legal cases of any Indian judicial organisation. The case data from the Indian Central Information Commission database was scrapped. Extracted case facts from case documents were used to predict the violation of the RTI Act. The problem of prediction has been defined as the binary classification problem. Three feature representation methods and created six learning models were used to compare and analyse the performance of the methods and models. The best performing model in the study was the linear kernel of SVM with Bag of words (with TF-IDF values) features. It achieved the accuracy of 70%

nearly 4% better than the second-best Sentence-BERT with Linear Kernel of SVM model.

It has been also observed that the violation cases have better predictive accuracy than the non-violation cases. The top 15 positive and negative weight features also throw light into the words, which were important in predicting a case outcome. Ultimately, it is believed that this study shall act like a catalyst for researchers and industry leaders in India and abroad to unlock the potential of Artificial intelligence in the field of law. Database created by the study can help researchers further develop their learning models and improve on the performance of this study. One of the major challenges to studies like this was labeling of the data to use in the learning model. Labeling requires domain expertise. Domain expertise can be costly and can act as a limitation to studies like this. Most of the case documents generated by the Indian legal organisations have no defined structure which makes it hard to extract case facts from them was among others the major challenge. It has also been understood that there is an immense room for the improvement and future work in the study. Artificial Neural Networks can improve the performance of the model. Also, Large datasets can help improve performance drastically. Frameworks like this can one day act as a y tool used by the judges to fasten the process of justice and lawyers to give quality services to their clients.

References

1. Centre for Law and Policy Research: Statistics Pack 2018 Supreme Court Observer, July 2019. https://clpr.org.in/wp-content/uploads/2019/07/Supreme-Court-Statistics-Pack.pdf (2019) [Online]. Accessed 04 Dec 2019
2. Central Information Commission: Uploaded decisions. https://dsscic.nic.in/cause-list-report-web/view-decision/ (2020) [Online]. Accessed 20 Sept 2020
3. Central Information Commission: Second appeal. https://cic.gov.in/second-appeal (2020) [Online]. Accessed 20 Sept 2020
4. Douglas, W.O.: Stare decisis. Colum. L. Rev. **49**(6), 735–758 (1949)
5. Martin, A.D., Quinn, K.M., Ruger, T.W., Kim, P.T.: Competing approaches to predicting supreme court decision making. Perspect. Polit. **2**(4), 761–767 (2004)
6. Brüninghaus, S., Ashley, K.D.: Generating legal arguments and predictions from case texts. In: Proceedings of the 10th International Conference on Artificial Intelligence and Law, pp. 65–74. ACM (2005)
7. Aletras, N., Tsarapatsanis, D., Preoţiuc-Pietro, D., Lampos, V.: Predicting judicial decisions of the European court of human rights: a natural language processing perspective. PeerJ Comput. Sci. **2**, e93 (2016)
8. Sulea, O.M., Zampieri, M., Vela, M., Van Genabith, J.: Predicting the law area and decisions of French supreme court cases. arXiv preprint arXiv:1708.01681 (2017)
9. Virtucio, M.B.L., Aborot, J.A., Abonita, J.K.C., Aviñante, R.S., Copino, R.J.B., Neverida, M.P., Osiana, V.O., Peramo, E.C., Syjuco, J.G., Tan, G.B.A.: Predicting decisions of the Philippine Supreme Court using natural language processing and machine learning. In: 2018 IEEE 42nd Annual Computer Software and Applications Conference (COMPSAC), vol. 2, pp. 130–135. IEEE (2018)
10. Bansal, N., Sharma, A., Singh, R.: A review on the application of deep learning in legal domain. In: IFIP International Conference on Artificial Intelligence Applications and Innovations, pp. 374–381. Springer (2019)

11. Xiao, C., Zhong, H., Guo, Z., Tu, C., Liu, Z., Sun, M., Feng, Y., Han, X., Hu, Z., Wang, H., et al.: CAIL2018: a large-scale legal dataset for judgment prediction. arXiv preprint arXiv:1807.02478 (2018)
12. Kim, Y.: Convolutional neural networks for sentence classification. CoRR, vol. abs/1408.5882 (2014)
13. Yan, G., Li, Y., Shen, S., Zhang, S., Liu, J.: Law article prediction based on deep learning. In: 2019 IEEE 19th International Conference on Software Quality, Reliability and Security Companion (QRS-C), pp. 281–284. IEEE (2019)
14. Ministry of Law and Justice. (Legislative Department): The right to information act, 2005. https://rti.gov.in/rti-act.pdf (2005). [Online]. Accessed 19 Oct 2020
15. The Pandas Development Team: Pandas-dev/pandas: pandas, Feb 2020
16. Ling, C.X., Sheng, V.S.: Class Imbalance Problem, pp. 171–171. Springer US, Boston, MA (2010)
17. Salton, G., Wong, A., Yang, C.-S.: A vector space model for automatic indexing. Commun. ACM **18**(11), 613–620 (1975)
18. Sammut, C., Webb, G.I. (eds.): TF–IDF, pp. 986–987. Springer US, Boston, MA (2010)
19. Mikolov, T., Chen, K., Corrado, G., Dean, J.: Efficient estimation of word representations in vector space. arXiv preprint arXiv:1301.3781 (2013)
20. Reimers, N., Gurevych, I.: Sentence-BERT: sentence embeddings using Siamese BERT-networks. In: Proceedings of the 2019 Conference on Empirical Methods in Natural Language Processing. Association for Computational Linguistics (2019)
21. Cristianini, N., Ricci, E.: Support Vector Machines, pp. 928–932. Springer US, Boston, MA (2008)
22. Webb, G.I.: Naïve Bayes, pp. 713–714. Springer US, Boston, MA (2010)
23. Laaksonen, J., Oja, E.: Classification with learning k-nearest neighbors. In: Proceedings of International Conference on Neural Networks (ICNN'96), vol. 3, pp. 1480–1483 (1996)
24. Fürnkranz, J.: Decision Tree, pp. 263–267. Springer US, Boston, MA (2010)
25. Ting, K.M.: Confusion Matrix, p. 260. Springer US, Boston, MA (2017)

On the Implementation of a Semantic Model for Intelligent Vehicle Navigation

Alessandro Correa Victorino and Marie-Hélène Abel

Abstract This paper considers the problem related to the navigation of multi-actuated intelligent vehicles in interaction with its occupants and its environment, by modeling and describing the semantics of the contexts associated to the movement. That is, formally defining the semantic model of the navigation context, as obstacles, pedestrians, roundabouts, information on the vehicle itself (load, power, etc.). This semantic modeling generates an additional degree of abstraction by "reasoning" about the information collected by the various sensors of an autonomous vehicle in order to build a navigation context for the latter. This navigation context is able to provide decision support for the autonomous or semi-autonomous vehicle (interaction to the driver). In this paper, we show initial modeling results by defining and selecting the relevant information in the creation of a context for the vehicle, structuring the contextual information so as to be able to reasoning on it and thus dynamically update the context in which the vehicle operates; maintaining information by choosing suitable data persistence models in order to be able to query previous simulations efficiently.

Keywords Semantic navigation · Context modelling · Autonomous navigation

1 Introduction

The safe navigation of an autonomous robotic vehicle performing tasks in an unknown environment requires perceiving the nearby environment, interacting with any spatial and temporal changes in this environment, and rescheduling its mission according to this dynamic. A first required functionality is the autonomous realization of movements, avoiding obstacles, in an a priori unknown environment.

Current state of works in autonomous navigation for robotic vehicles considers the displacement of the vehicle in a local context (road monitoring, obstacle avoidance,

A. C. Victorino (✉) · M.-H. Abel
Sorbonne universités, Université de Technologie de Compiègne, CNRS, Heudiasyc UMR 7253, CS 60319, 60203 Compiègne Cedex, France
e-mail: alessandro.victorino@hds.utc.fr

L. Troiano et al. (eds.), *Advances in Deep Learning, Artificial Intelligence and Robotics*, Lecture Notes in Networks and Systems 249,
https://doi.org/10.1007/978-3-030-85365-5_5

etc.) and based on the perception of the environment. In [1] a navigation method is presented by combining a visual-based navigation method with an obstacle avoidance method based on the methodology of dynamic window approach. Other works consider the global navigation problem of running a complete itinerary with the management of intersections and roundabouts [2, 3]. These navigation methodologies enable the autonomous vehicle to detect obstacles in an "empirical" manner via various sensors without however "understanding" the nature of these obstacles. Building a semantic modeling would thus make it possible to generate an additional degree of abstraction by "reasoning" about the information collected by the various sensors of an autonomous vehicle in order to build a navigation context for the latter. Ultimately, this navigation context should be able to provide decision support for the autonomous or semi-autonomous vehicle (to inform the driver).

Context is a complex notion [4]. According to [5], it is the key element used to help intelligent entities understand how events in the surrounding world influence their behavior. It is very poorly defined, the definitions in the literature are too dependent on their own contexts [6]. The latter depend on the field of application [7], for example:

- The context is the set of circumstances surrounding an event or an object [6]. This definition is widely used in the field of psychology.
- Context is a set of explicit variables that models contextual factors in the underlying domain, for example, time, place, environment, devices, occasion, etc. [8]. This definition comes from recommendation systems.
- The context is the set of environmental states and parameters that determine the behavior of an application or in which an event occurs and is of interest to the user [9]. This definition is generally used in the field of contextual computing.

Note also that in the field of psychology, the notion of context is often used in the following sense: set of situational elements in which the object being processed is included [10].

The basic problem treated in our research is to combine these methods by taking into account a semantic model of the navigation context considering for example obstacles, pedestrians, roundabouts, crossings, information on the vehicle itself (load, power, etc.). On the one hand, it is a question of formally defining these elements of the navigation context that we can obtain, for example from the visual perception of the road and the elements around the vehicle provided by an on-board camera or through inter-vehicles communication, and on the other hand show their use in a decision support process, for example in the context of an overtaking task, emergency stop and entering and exiting roundabouts. The on-board system will thus navigate by taking into account the semantic representations of the context (itself + environment) in which it operates.

In this paper we present initial results of the semantical context of an intelligent vehicle navigation, by defining and selecting the relevant information in the creation of a context for the vehicle, structuring the contextual information so as to be able to reasoning on it and thus dynamically update the context in which the vehicle operates, and choosing suitable data persistence models in order to be able to maintain information and querying previous simulations efficiently.

2 Context Definition

In order to develop a semantic context model, it is necessary to identify the elements that constitute the context associated with displacement of a vehicle.

We are interested in the spatial context. The latter can be divided into three dimensions:

- The static spatial context: information that can have an impact on the user's environment;
- The dynamic spatial context: georeferenced information collected by different types of sensors;
- The internal spatial context: information accessible from local devices such as GPS.

In the case of intelligent vehicles, it also seems relevant to integrate a temporality into the system to define periods of crowds, for example. These temporal data coupled with spatial data would allow the vehicle to predict the density of traffic that it would encounter in order to adapt its driving accordingly. This temporality of the context was not taken into account in our model but it could be taken into account, in the long term, when the driver would indicate his route on the application linked to the driving assistant.

2.1 Context Modeling

The simulated navigation scenarios are generated by a python script. We decided to categorize our context in the navigation scenarios as follows, as shown in Fig. 1.

- Perennial information (StaticInfo class on the UML in Fig. 1): everything relating to the conditions in which the trip will be made. This information can be entered upstream of the journey and may concern:

  –passengers: the number, presence of elderly people, children in infants or pregnant women, etc.;
  –the vehicle: type, model, technical performance, etc.;

- Real-time information retrieved from vehicle sensors:

  –the current speed of the vehicle;
  –weather conditions;
  –the type of road taken at a time t and the associated authorized speed;

- The types of obstacles that the vehicle encounters at time t:

  –pedestrian crossing;
  –Railroad crossing.

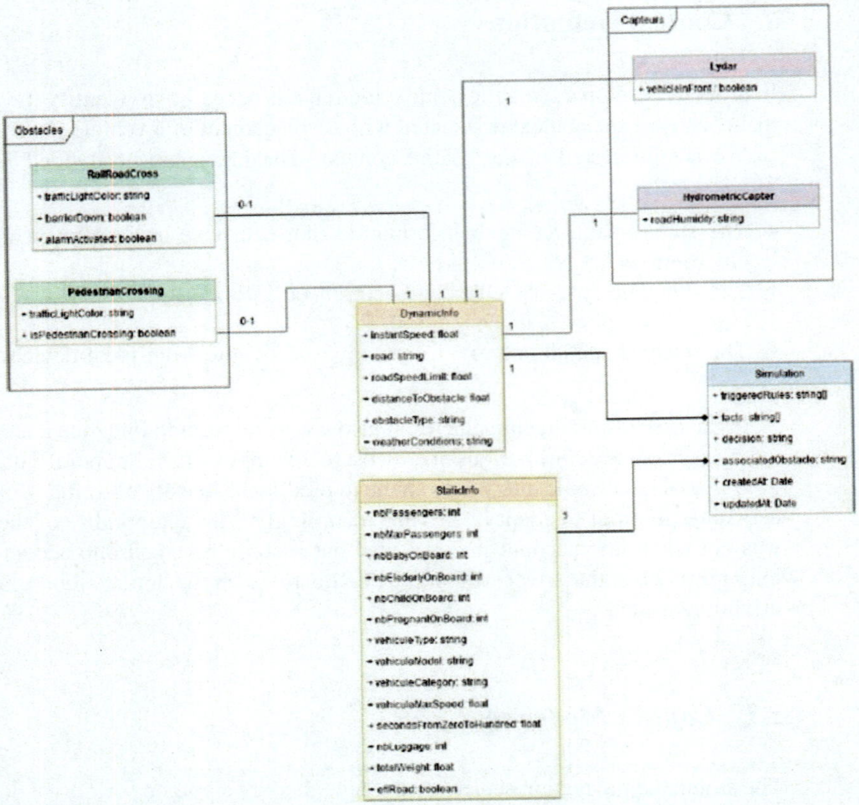

Fig. 1 Persistency of the context

In the present case, the elements are limited to these listed above, but they could be completed.

2.2 *Development and Technical Implementation*

Choice of the technologies The technological choices to develop a version of the POC were guided by two main elements (as shown in Fig. 2).

- *On the server side*, the desire to be able to efficiently query the data persisted in the database in order to easily retrieve contextual information on previous modeled simulations as well as on the vehicle's final decision-making. We therefore build the back-end on the model of an API service that could be queried from the outside

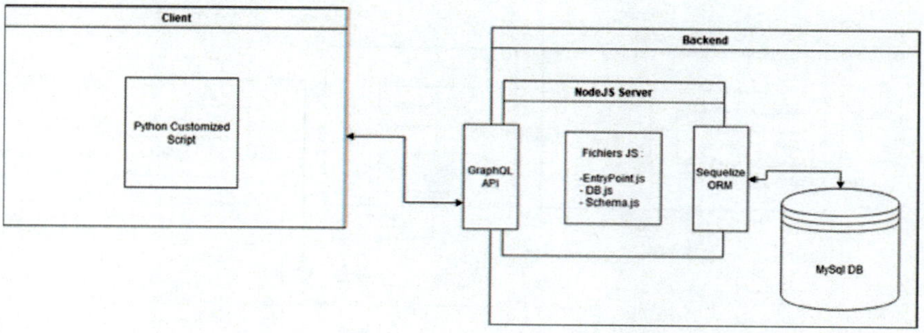

Fig. 2 Global structure of the model

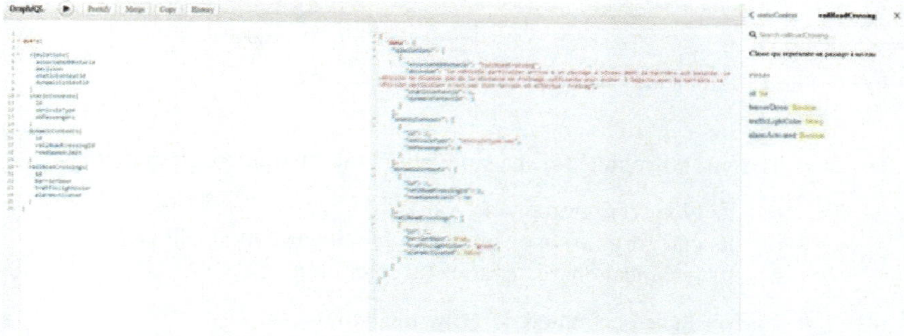

Fig. 3 Example of the GraphQL interface for the query to the API

to access the data. This was done using by coupling GraphQL[1] with a Node.js server given the documentation available regarding this association. Figure 3 shows the advantage of this model in retrieving data from multiple tables in a single query. Regarding the general system of data basis (GSDB), we favored a relational approach by using MySql.

- *On the client side*, we therefore implemented the simulation generation script in python with an object-oriented programming paradigm. To tie the object-oriented programming paradigm of our script with data persistence in the MySql GSDB, we resorted to an object-relational mapping (ORM): *Sequelize*, which is the most widely documented and preferred ORM for node.js.

Technical implementation The development and technical implementation of the semantical navigation model is presented in a schematic way in Fig. 4.

Scenario simulation with Python Each new simulation is characterized by a set of variable parameters to allow the definition of a unique context. These randomly generated contextual parameters concern, among others:

[1] Strongly typed query language developed by Facebook and under open-source license since 2015.

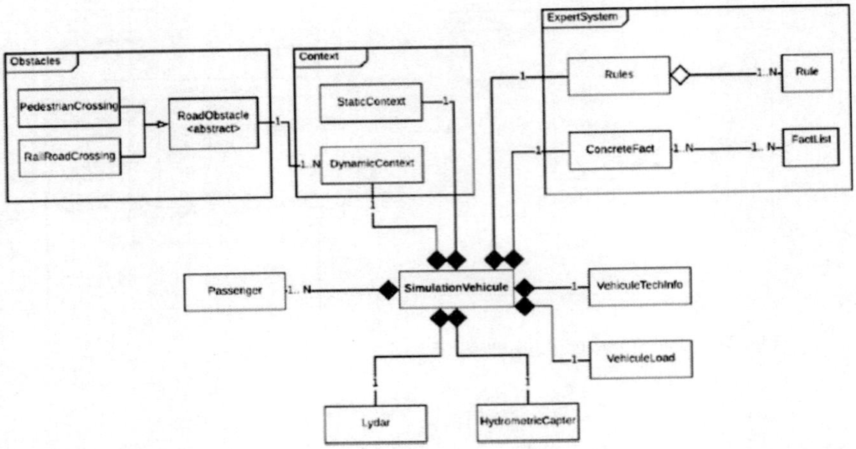

Fig. 4 UML of the global application

- The virtual vehicle for the simulation, taking into consideration:
  - o Its type (car, motorcycle, van);
  - o Its category (private vehicle, public intervention vehicle);
  - o Its technical characteristics (acceleration, speed).
- The obstacle encountered during the simulation.
  - o Each obstacle corresponds to a class with its own characteristics (presence of a traffic light, barrier, sound signal, physical retarder on the ground, delimiter, etc.).
- The number of vehicle passengers with certain specific detection criteria:
  - o The age of each passenger for the elderly or frail infants;
  - o Potential illness or injury;
  - o Possibility for a female passenger to be pregnant.
- The load plus total weight of the vehicle:
  - o Sensors associated with the vehicle during the simulation;
  - o Hygrometry sensor;
  - o Instant vehicle speed;
  - o Lidar.

Although the data is generated randomly, it is obviously subject to verification so as not to face certain absurd situations (pregnant female passenger, number of motorcycle passengers greater than 2, etc.). We start by randomly generating a new data set for each simulation in order to build a new context for each simulation. The information generated randomly by the python client script (see Fig. 2) is specified to the user at the start of each new simulation. It is this information that we will persist in our database.

In addition to these perception data, an expert system (see Fig. 3) which will propose an action for the driver to perform according to the obtained simulation. The reasoning leading to this decision making will also be persisted.

3 Expert System Implementation

An expert system is a tool capable of reproducing the cognitive mechanisms of an expert, in a particular field. It consists of three parts (see Fig. 3):

- An event basis: The information available before the intervention of the inference engine;
- A rule basis: A set of rules resulting from the expertise;
- An inference engine: That confronts the rules with the facts, allowing the expert system to conduct reasoning to dynamically deduce new facts from those already available.

The inference engine implements the following formal logic:

- Order 0: A fact is a proposition, true or false;
- 0+ order: A fact is a value;
- Order 1: A fact is a variable.

In our case, the engine is type 0 with forward chaining, that is to say, we start from the premises to draw new conclusions. These findings enrich the knowledge base and may become the premises for other rules.

A specific event is trigged based on a basis of rules.

4 Conclusions

We presented in this paper an idea in order to modelling the semantical context around a vehicle navigation. The presented semantical model uses the simulator data to identify the elements concerning the vehicle navigation at a given instant and to persist this information in order to be able to highlight the relevant information capable of identifying an obstacle. The simulation focused on obtaining different configurations of the same obstacles to generate contextual information relating to this obstacle (color of the traffic lights, positioning of a level crossing barrier, presence or not of a pedestrian crossing for a pedestrian crossing, etc.). An expert system capable of reproducing the cognitive mechanisms of an expert, was modelled. Based on set of events, a set of rules, the inference engine confronts the rules with the facts, in order to deduce new facts from those already available.

References

1. Lima, D.A., Victorino, A.C.: A hybrid controller for vision-based navigation of autonomous vehicles in urban environments. IEEE Trans. Intell. Transp. Syst. **17**(8), 2310–2323 (2016)
2. Pereira, E., Lima, D.A., Victorino, A.: Autonomous vehicle global navigation approach associating sensor based control and digital maps. IEEE International Conference on Robotics and Biomimetics (ROBIO'2014), Bali, Indonesia (2014)
3. Kenmogne, I.-F., Lima, D.A., Victorino, A.C.: A global navigation management architecture applied to autonomous robots in Urban environments. In: 18th International IEEE Conference on Intelligent Transportation Systems (ITSC'15), Canary Islands, Spain (2015)
4. Adomavicius, G., Jannach, D.: Preface to the special issue on context-aware recommender systems. User Model. User Adapt. Interact. **24**, 1–5 (2013)
5. Kofod-Petersen, A., Cassens, J.: Using activity theory to model context awareness. In: Modeling and Retrieval of Context, pp. 1–17. Springer, Berlin, Heidelberg (2006)
6. Bazire, M., Brézillon, P.: Understanding context before using it. In: International and Interdisciplinary Conference on Modeling and Using Context, pp. 29–40. Springer, Berlin, Heidelberg (2005)
7. Palmisano, C., Tuzhilin, A., Gorgoglione, M.: Using context to improve predictive modeling of customers in personalization applications. IEEE Trans. Knowl. Data Eng. **20**(11), 1535–1549 (2008)
8. Adomavicius, G., Tuzhilin, A.: Context-aware recommender systems. In: Recommender Systems Handbook, p. 217 (2011)
9. Chen, G., Kotz, D.A.: Survey of context-aware mobile computing research. Technical Report, Dartmouth College, Hanover, NH, USA (2000)
10. Bastien, C.: Contexte et situation. In: Houdé, O., Kayser, D., Koenig, O., Proust, J., Rastier, F. (eds.) Dictionnaire des Sciences Cognitives, PUF, Paris (1998)

Twenty Years of Assessment of Drought Conditions Over The Gangapur Region of Aurangabad District in Maharashtra (India) Using Remote Sensing Indices

Rashmi R. Nitwane, Vaishali D. Bhagile, and R. R. Deshmukh

Abstract One of the nature's worst calamity is Drought. Any type of Drought affects the economy of the country, whether it is agricultural drought, meteorological or hydrological drought. In India there are certain regions which face droughts continually, one such region is Marathwada region of Maharashtra which is continuously reeling from agricultural drought from last few decades. Gangapur region located on the west of Aurangabad is a Taluka place which covers total land area 130,846 ha. In this paper we have analyzed the drought conditions over the Gangapur region of Aurangabad district from last two decades i.e. from May 2000 to Apr 2020. The drought analysis is done for the two major cropping cycles in India namely Kharif and Rabi. The multi temporal analysis using MODIS based Vegetation Condition Index (VCI) is done to find the drought onset period, duration of drought and severity of drought. We found that VCI proves to be useful in detecting the drought conditions for both the crop cycles efficiently.

Keywords Vegetation condition index · Drought · Temporal analysis · Machine learning · Remote sensing

R. R. Nitwane · V. D. Bhagile · R. R. Deshmukh (✉)
Department of Computer Science and Information Technology, Dr. B.A.M. University, Aurangabad, Maharashtra, India
e-mail: rrdeshmukh.csit@bamu.ac.in

L. Troiano et al. (eds.), *Advances in Deep Learning, Artificial Intelligence and Robotics*, Lecture Notes in Networks and Systems 249,
https://doi.org/10.1007/978-3-030-85365-5_6

1 Introduction

Agricultural Droughts refers to deficit in soil moisture during the crop growth stages. When there is crop failure at any stage of crop growth cycle it results in to economic loss for farmers as well as it adversely affects the economic growth of the country. Indian economy is largely agricultural and agriproducts based and therefore repeated drought situation in any part of the country is devastating. There are three major cropping seasons in India they are Kharif, Rabi and Zaid. The Kharif season crops are usually rainfed (dependent on precipitation) whereas Rabi crops cultivation is limited to land with irrigation facilities. In India the largest cultivation is of Kharif crops. These crops are sown during monsoon period and sowing start in the last week of May or first week of June every year The corresponding Day of Year (DOY) is May last week (145/146) and the harvesting is done in last week of September the corresponding DOY (273/274), whereas Rabi crops sowing period starts in October every year corresponding DOY (289) and harvested in the month of January. The Marathwada region of Maharashtra has suffered from drought situation during last several decades. In this paper we have taken Gangapur region of Aurangabad district of Marathwada region Maharashtra as study region for analysis of the drought situation. In the Modis13Q1 dataset we have taken the corresponding day of year for sowing and harvesting of kharif as (145 and 273) and for Rabi(289 and 017).

1.1 Literature Review

In the context of drought monitoring, assessment and prediction various researchers have developed and are in the process of developing drought indices. These indices are segregated in to three categories these are meteorological, agricultural and hydrological drought indices [1]. In addition to these Indices there are also some researchers who have developed integrated drought indices with the help of Remote sensing and machine learning [2]. Peng Sun et al. developed Integrated drought monitoring model using multisource data such as meteorological, soil moisture and remote sensing data while [3]. Pesaresi et al. developed symbolic machine learning model for analyzing big remote sensing data. It has been observed that remote sensing data have large spatial coverage for drought monitoring, the Table 1 lists some of the remote sensing drought indices developed in more than three decades.

In this study we have evaluated the ability of Vegetation condition Index to analyze the drought impact over the region of Gangapur, during the last two decades i.e. 2000–2020.

Table 1 Some drought indices derived from remote sensing

Drought Index	Component	Remote Sensing Data	References
VCI	NDVI/EVI	MOD13Q1	Kogan [4]
TCI	LST	MOD11A2	Kogan [4–6]
TVDI	NDVI/LST	MOD09A1/MOD09A2	Sandholt et al. [7]
VTCI	NDVI/LST	MODIS data	Sun et al. [8]
PDI	NIR,RED	MODIS data	Ghulam et al [9]
MPDI	NIR,RED	ETM + , MODIS data	Ghulam et al. [9]

2 Data and Method

2.1 Study Area

The study area of Gangapur is a taluka place in district Aurangabad, Maharashtra which is located on the west side of Aurangabad—Ahmednagar highway. The geographical location of the study area is 19.699 N and 75.010E. The total land area of Gangapur is 1,30,846 Hectares out of which the total agricultural area is approx. 94,298 hectares. The government agencies declared 2006, 2007, 2012 as severe drought years in agricultural census [10, 11]. The major crops grown in the area are wheat, cotton, bajra etc. Due to rainfall variability over the region Marathwada is sensitive to droughts. Severe as well as mild droughts has affected the quality of the crop yield over the years. The image of Gangapur area of Aurangabad District Maharashtra is shown below (Fig. 1).

2.2 Data

Didan [12] the remote sensing data used in this study is Moderate Resolution Imaging Spectroradiometer 16 day MODIS Vegetation Index (VI) product (EVI) provided at 250 m resolution for the time range May 2000- Apr 2020.The MODIS VI product (MOD13 Q1)provide consistent spatial, temporal time series comparisons of global vegetation conditions [6, 13]. The MODIS Vegetation Index product Enhanced Vegetation Index (EVI) has improved sensitivity over high biomass regions and improved vegetation monitoring capability through a decoupling of the canopy background signal and reduction in atmosphere influences. The EVI incorporates atmospheric resistance concept along with the removal of soil brightness induced variations in Vegetation Index based on SAVI. To address the issue of erratic behavior of EVI over bright targets the 3 band EVI is replaced with 2 band EVI backup algorithm which does not use blue band. The formula for 3-band EVI is given as

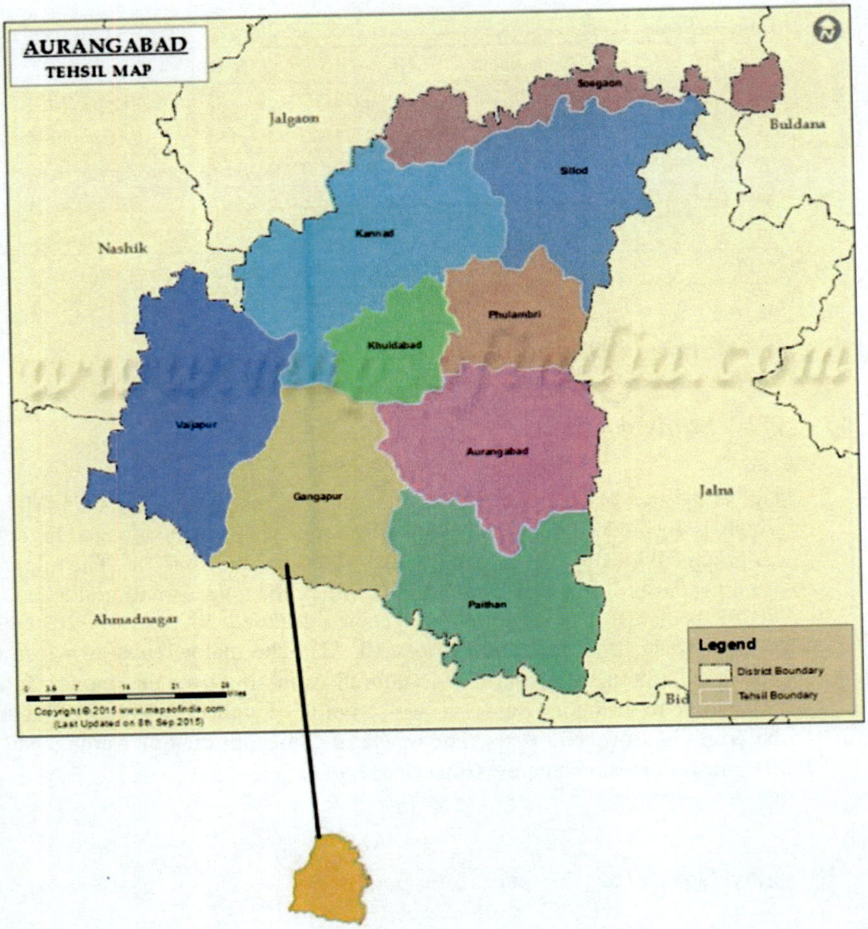

Fig. 1 Gangapur region (Aurangabad District) Source Z.P Aurangabad, Maharashtra

$$EvI = G\frac{NIR - RED}{NIR + C_1 RED - C_2 Blue + L} \tag{1}$$

where, NIR, RED and Blue are fully or partially atmospheric corrected surface reflectance, L is canopy Background adjustment and C1 & C2 are coefficients of aerosol resistance terms and G is the gain or scaling factor.

And EVI 2-band formula is

$$EvI = 2 \cdot 5\frac{NIR - RED}{NI\mathbb{R} + 2 \cdot 4RED^{+1}} \tag{2}$$

2.3 Methods

In this study we have implemented VCI on the time series dataset. The Vegetation Condition Index (VCI) proposed by Kogan [6, 14] has proven its usefulness for drought detection and tracking in numerous cases. Validations showed that the VCI has excellent ability to detect drought and to measure time of its onset, intensity, duration, and impact on vegetation. The VCI provides accurate drought information not only for well-defined, prolonged, widespread, and intensive droughts, but also for very localized, short-term, and not so well-defined droughts. The VCI implementation on EVI product of MODIS is done following the workflow diagram as given by [15–17] shown as below in Fig. 2. The whole process of data preparation, layer stacking and preprocessing is conducted using R studio software. The brief description of VCI is given in next section.

2.4 Vegetation Condition Index

Vegetation Condition Index was first proposed by [4, 6] and is currently operational in Iran. VCI is a proxy for moisture condition and it focusses on impact of drought on the vegetation. It was originally developed by scaling NDVI values from 0 to 1 using minimum and maximum NDVI values calculated from long term record although VCI suitably reflects the spatial distribution and development of drought over the region but in some cases VCI based on NDVI does not perform well but VCI based on EVI dataset (MOD13 Q1) shows some advantages as listed below.

- There is no distortion in the reflected light caused by the particles in the air or by ground cover below the vegetation.
- The EVI data product does not become saturated as the NDVI as it happens when viewing rainforests and large amounts of chlorophyll.

The VCI is calculated as

$$\frac{\mathrm{VCI} = \mathrm{EVI_C} - \mathrm{EVI_{Min}}}{\mathrm{EVI_{Max}} - \mathrm{EVI_{Min}}} * 100 \tag{3}$$

where,

EVI_C = EVI * Cloudmask.

i.e. a single layer stack with applied cloud mask.

EVI_Max = Calculating max value for layer stack for each pixel.

EVI_Min = Calculating min value for layer stack for each pixel.

The VCI Value thus calculated gives the percent of drought for a given month. The VCI value close to 0–35% shows severe drought condition whereas the VCI percent above 50% shows normal to above normal condition for vegetation.

In this study the data is with 16 days intervals with each day of the year represented as 001–353 and the layer stacks are formed according to the number of days. The

Fig. 2 Workflow of VCI in R (UNSPIDER Knowledge portal)

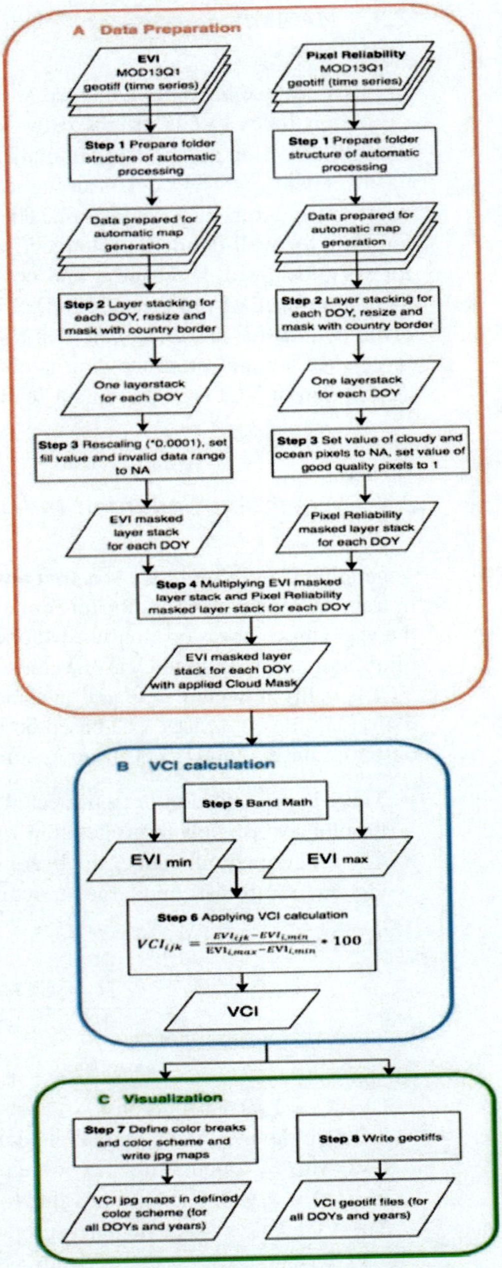

EVI Geotiff files and Pixel reliability geotiff files are processed simultaneously and after the layer stacking rescaling is done by (0.0001). Next the EVI masked layer stack for each Day of year (DOY) is calculated. Then EVI masked layer stack and pixel reliability masked layer stack for each day of year is done to finally calculate the VCI. The results are shown in the form of maps generated after execution of code and further analysis is done. The Maps generated are shown on the next page.

3 Results and Discussion

The Kharif crops sowing cycle begins towards the end of May and beginning of June every year in India The Modis dataset (MOD13Q1) contains total 26 Day Of Year (DOY) starting from 001–353 with 16 day interval between them each DOY has 20 samples starting from year May 2000 to April 2020. To maintain the uniformity for analysis purpose we have taken DOY 145 as the Kharif sowing start date and DOY 273 as Kharif harvesting start date. Similarly, we have taken DOY 289 as Rabi sowing start date and DOY 017 as Rabi harvesting start date. The VCI calculation shows that the years 2001, 2004 and 2016 were having severe drought conditions in the beginning of kharif sowing cycle moreover the years 2001, 2002, 2015, 2018 showed severity in drought conditions over the region towards the harvesting start date which clearly indicates the lack of appropriate monsoon precipitation over the region during the crop growth cycle. Even if 2004 and 2016 didn't show severe drought towards the harvesting cycle the crop yield is affected due to initial delay of monsoon and drought like conditions. The overall Kharif crops over these two decades were severely affected in the years 2001,2002,2003,2004 and moderately affected in the years 2017 and 2018. The beginning of Rabi crops starts in October every year in Maharashtra the corresponding DOY 289 as shown below proves that 2003, 2004, 2018 were drought year through out both the seasons and 2013 and 2016 were moderate drought years and the remaining years showed partial drought conditions through out the year. The results showed that 2019 was good for crops both Kharif and Rabi crops. It is necessary to test these results for prediction too and comprehensive study should be made for proper prediction strategy.

Kharif Crop Season Cycle Sowing Period Day of Year 145 (May) For 20 Years (2000–2019)

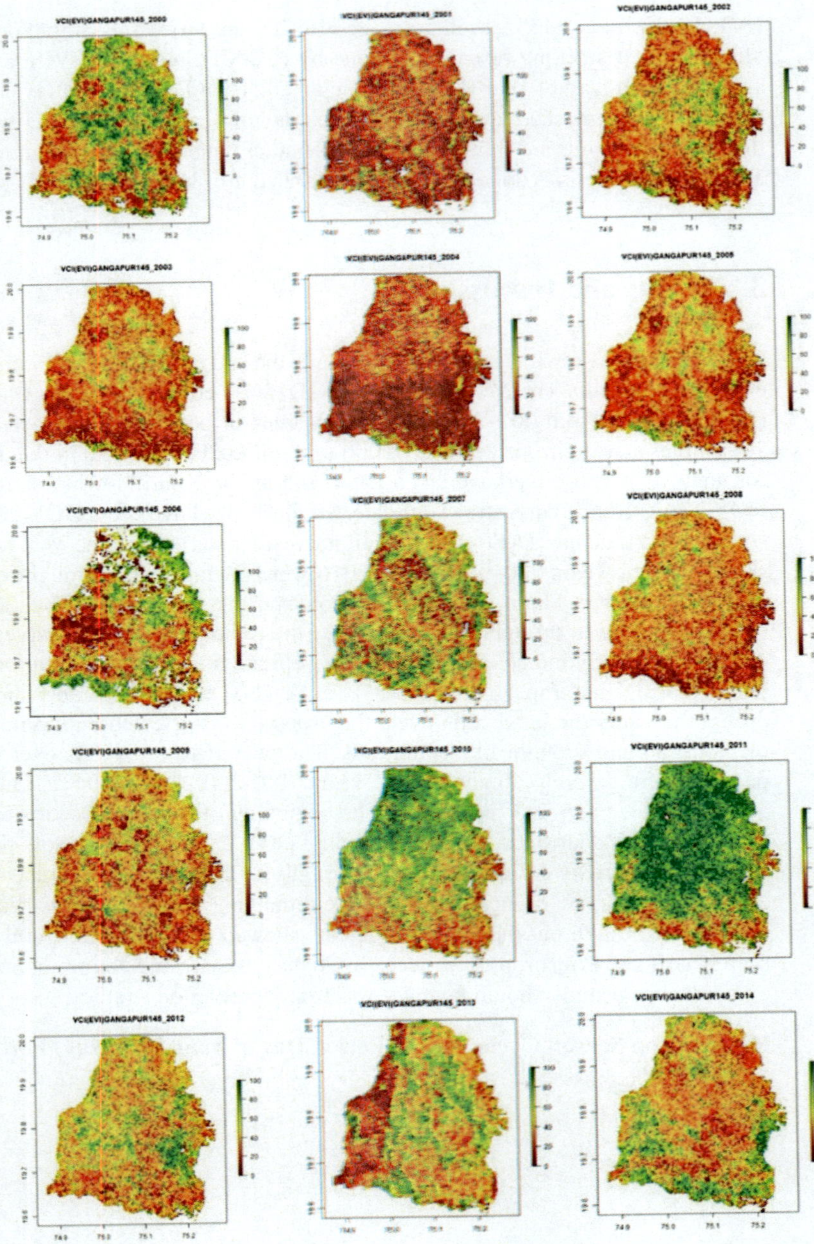
VCI(EVI)GANGAPUR145_2000
VCI(EVI)GANGAPUR145_2001
VCI(EVI)GANGAPUR145_2002
VCI(EVI)GANGAPUR145_2003
VCI(EVI)GANGAPUR145_2004
VCI(EVI)GANGAPUR145_2005
VCI(EVI)GANGAPUR145_2006
VCI(EVI)GANGAPUR145_2007
VCI(EVI)GANGAPUR145_2008
VCI(EVI)GANGAPUR145_2009
VCI(EVI)GANGAPUR145_2010
VCI(EVI)GANGAPUR145_2011
VCI(EVI)GANGAPUR145_2012
VCI(EVI)GANGAPUR145_2013
VCI(EVI)GANGAPUR145_2014

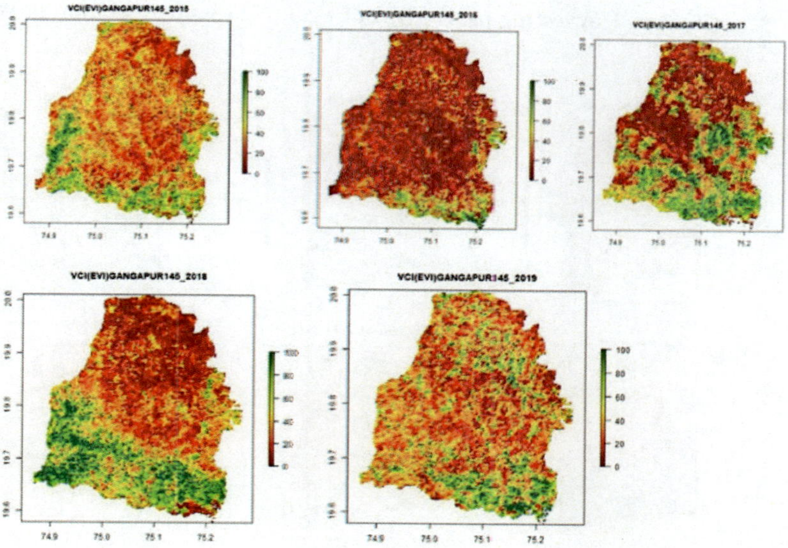
VCI(EVI)GANGAPUR145_2015
VCI(EVI)GANGAPUR145_2018
VCI(EVI)GANGAPUR145_2019
74.9
75.0
75.1
75.2

Kharif crops Harvesting period Day of Year 273 (September) for 20 years 2000–2019

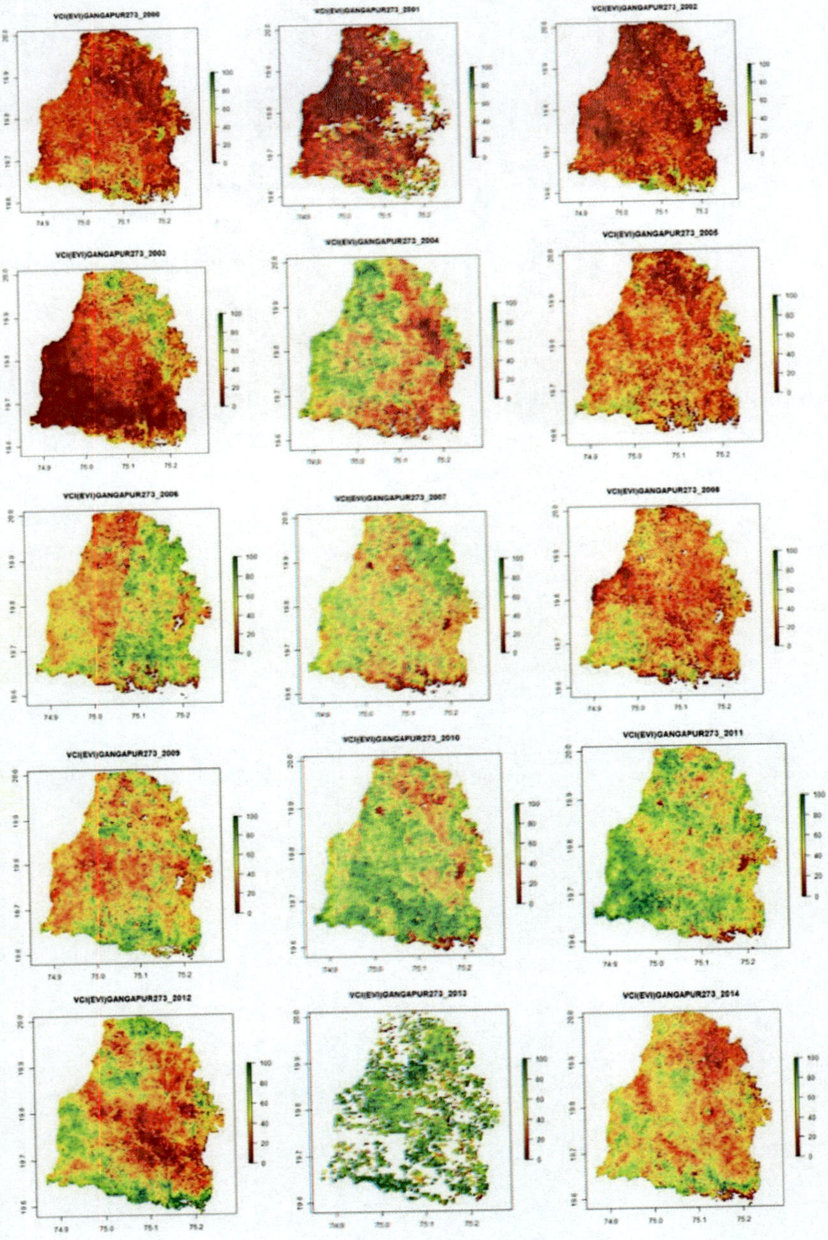

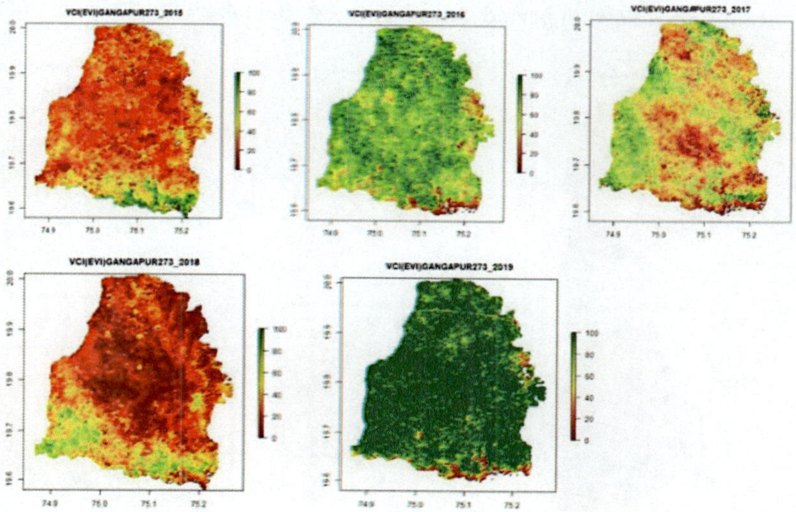
VCI(EVI)GANGAPUR273_2015
VCI(EVI)GANGAPUR273_2016
VCI(EVI)GANGAPUR273_2017
VCI(EVI)GANGAPUR273_2018
VCI(EVI)GANGAPUR273_2019
74.9
75.0
75.1
75.2
19.6
19.7
19.8
19.9
20.0
0
20
40
60
80
100

Rabi Crops Sowing period Day of Year 289 (October) for 20 years 2000–2019

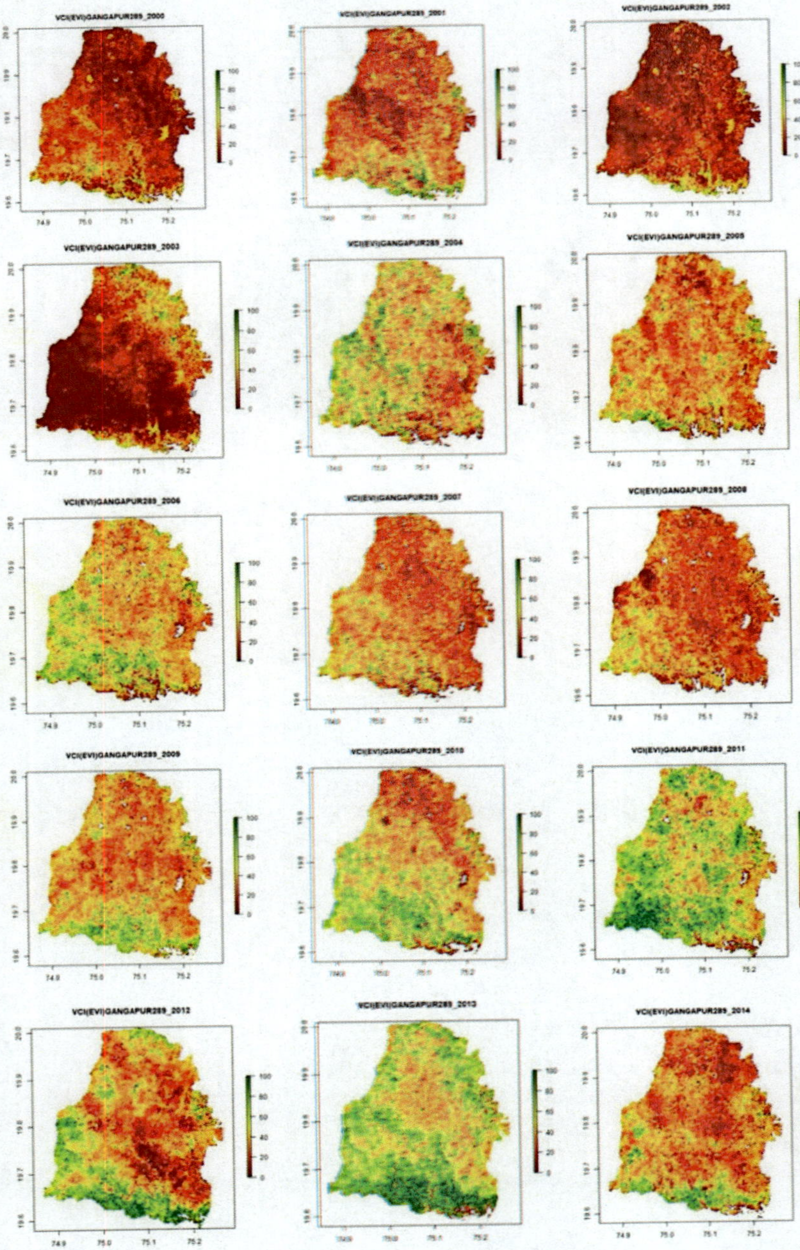

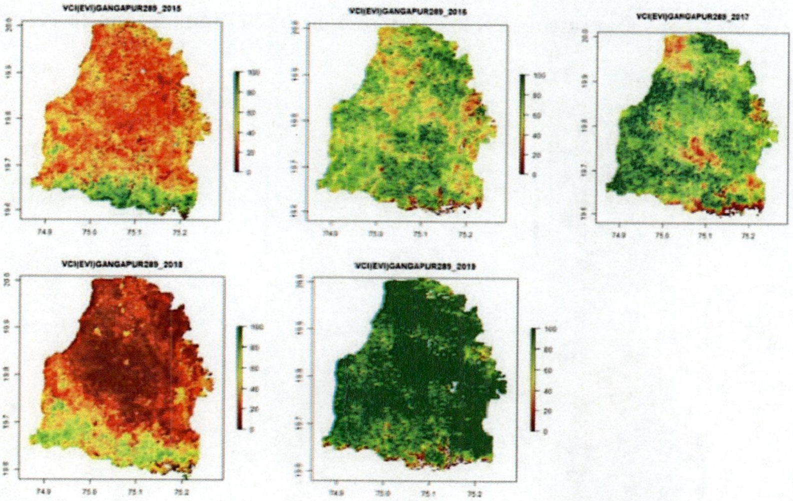
VCI(EVI)GANGAPUR289_2015
VCI(EVI)GANGAPUR289_2016
VCI(EVI)GANGAPUR289_2017
VCI(EVI)GANGAPUR289_2018
VCI(EVI)GANGAPUR289_2019

Rabi Crops Harvesting Period Day of Year 017 (January) for 20 years 2001–2020

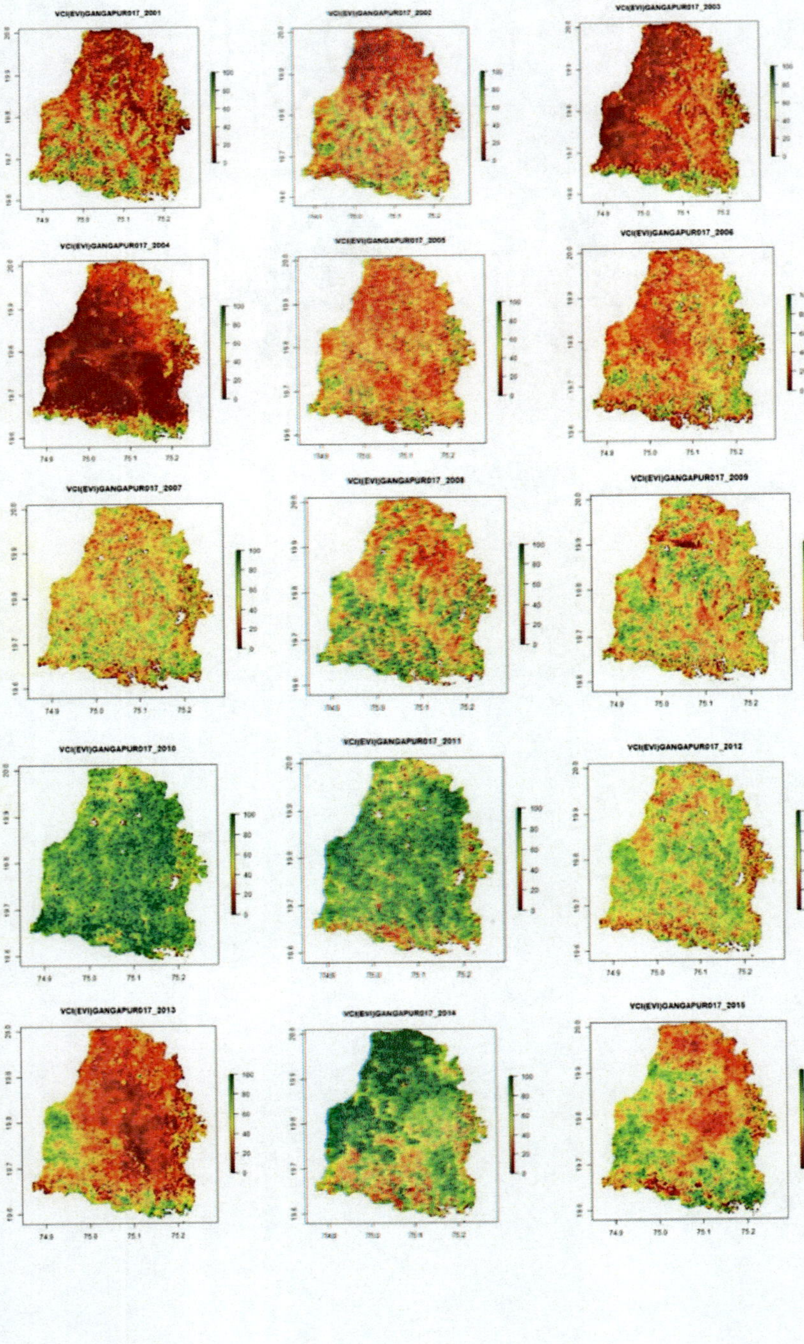

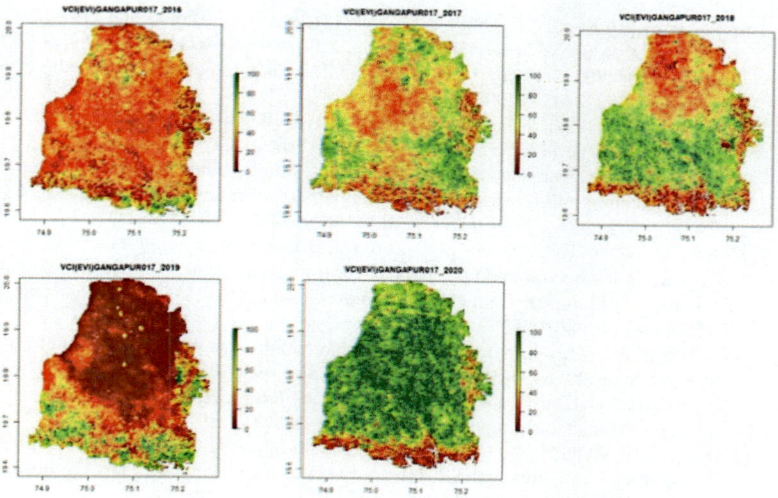

4 Conclusion

The results clearly signifies that times series analysis of the geographical area is very useful in understanding the changes in the crop cycle as well as giving an insight for building robust methods for drought prediction. Therefore it becomes necessary to test these results for prediction and do comprehensive study for proper prediction strategy. The future scope of this study is to build prediction algorithm based on the integration of remote sensing data and machine learning tools for forecasting the drought at taluka level.

References

1. Aswathi, P., Nikam, B., Chouksey, A., Aggarwal, S.: Assessment and monitoring of agricultural droughts in Maharashtra using meteorological and remote sensing based Indices. Isprs Ann. Photogramm. Remote Sensing Spatial Inf. Sci. IV(5), 253–264 (2018). https://doi.org/10.5194/isprs-annals-IV-5-253-2018
2. Sun, P., Zhang, Q., Wen, Q., Singh, V. P., Shi, P.: Multisource data-based integrated agricultural drought monitoring in the Huai River basin, China. J. Geophys. Res. Atmosph. **122**(10), 751–772. (2017). https://doi.org/10.1002/2017JD027186
3. Pesaresi, M., Syrris, V., Julea, A.: Analyzing Big Remote Sensing Data via Symbolic Machine Learning (2016).
4. Kogan, F.N.: Application of vegetation index and brightness temperature for drought detection , Adv. Space Res. **15**(11), 91–100. https://doi.org/10.1016/0273-1177(95)00079-T. ISSN0273-1177. http://www.sciencedirect.com/science/article/pii/027311779500079T

5. Seiler, R.A., Kogan, F., Sullivan, J.: AVHRR-based vegetation and temperature condition indices for drought detection in Argentina. Adv Space Res. **21**(3) 481–484. (1998).https://doi.org/10.1016/S0273-1177(97)00884. ISSN 0273-1177. http://www.sciencedirect.com/science/article/pii/S0273117797008843
6. Kogan, F.N.: Global drought watch from space. Bull. Amer. Meteor. Soc. **78**, 621–636 (1997). https://doi.org/10.1175/1520-0477(1997)078%3c0621:GDWFS%3e2.0.CO;2
7. Sandholt, I., Rasmussen, K., Andersen, J.: A simple interpretation of the surface temperature/vegetation index space for assessment of surface moisture status. Remote Sens. Environ. **79**, 213–224 (2002). https://doi.org/10.1016/S0034-4257(01)00274-7
8. Sun, W., Wang, P.-X., Zhang, S.-Y, Zhu, D.-H., Liu, J.-M., Chen, J.-H., Yang, H.-S.: Using the vegetation temperature condition index for time series drought occurrence monitoring in the Guanzhong Plain, PR China Int. J. Remote Sens. **29**(17–18), 5133–5144. https://doi.org/10.1080/01431160802036557
9. Ghulam, A., Qin, Q., Kusky, T.: A re-examination of perpendicular drought indices. Int. J. Remote Sens. **29**, 6037–6044 (2008). https://doi.org/10.1080/01431160802235811
10. http://krishi.maharashtra.gov.in/Site/Upload/Pdf/latur_cdap.pdf
11. http://krishi.maharashtra.gov.in/Site/Upload/Pdf/MSAAPC.pdf
12. Didan, K.: MOD13Q1 MODIS/Terra Vegetation Indices 16-Day L3 Global 250 m SIN Grid V006 [Data set : Aurangabad]. NASA EOSDIS Land Processes DAAC. https://modis.gsfc.nasa.gov/data/dataprod/mod13.php (2015)
13. MODIS VI User's Guide: http://vip.arizona.edu/MODIS_UsersGuide.php
14. Liang, L., Sun, Q., Luo, X., Wang, J., Zhang, L., Deng, M., Di, L., Liu, Z.: Long-term spatial and temporal variations of vegetative drought based on vegetation condition index in China. Ecosphere **8**(8), e01919 (2017). https://doi.org/10.1002/ecs2.1919
15. Jain, S.K., Keshri, R., Goswami, A., Sarkar, A., Chaudhry, A.: Identification of drought-vulnerable areas using NOAA AVHRR data. Int. J. Remote Sens. **30**(10), 2653–2668 (2009). https://doi.org/10.1080/01431160802555788
16. Thiruvengadachari, S., Gopalkrishna, H.R.: An integrated PC environment for assessment of drought. Int. J. Remote Sens. **14**(17), 3201–3208 (1993). https://doi.org/10.1080/01431169308904434
17. Du, M., Liu, Y.: Regional drought monitoring and analyzing using MODIS data—a case study in Yunnan Province. IFIP Adv. Inf. Commun. Technol. **345**, 243–251 (2010). https://doi.org/10.1007/978-3-642-18336-2_29

Educational Resources Recommender System for Teachers: Why and How?

Nader N. Nashed, Christine Lahoud, and Marie-Hélène Abel

Abstract In a world full of discrepancies, teachers usually perform their jobs under a huge amount of pressure and especially who work in severe conditions in the suburban and rural areas. The key for a solution that helps ease the hard-working conditions, is integrating proper educational resources along courses' contents. Moreover, the searching process for the aiding educational resources in a huge aggregation of unorganized data that is available through the Web could be a troublesome task to achieve. As a result, they follow a hard path to allocate the appropriate resources to facilitate the educational process. This research aims to define the importance of educational recommender systems (ERSs) to solve the teacher's problems and the methodology of recommending personalized educational resources for them; experienced or young, working in urban or rural areas, teaching elementary or university students.

Keywords Semantic web · Machine learning · Recommender systems · Education

1 Introduction

Teachers' struggles during the initial phase in their educational career can lead to failures or even the career's end at their very first step. Thus, the early detection and confrontation of these struggles is the core of the teacher's success and setting him on the right path [1]. The introduction of youth to the teaching career has a significant impact on their professional development for the upcoming years. We can define any teacher's working struggles through multiple aspects concerning the teacher himself

N. N. Nashed (✉) · C. Lahoud
Université Française d'Égypte, Cairo, Egypt
e-mail: nader.nashed@utc.fr

C. Lahoud
e-mail: christine.lahoud@ufe.edu.eg

N. N. Nashed · M.-H. Abel
HEUDIASYC, Université de Technologie de Compiègne, Sorbonne universités, Compiègne, France
e-mail: marie-helene.abel@utc.fr

L. Troiano et al. (eds.), *Advances in Deep Learning, Artificial Intelligence and Robotics*, Lecture Notes in Networks and Systems 249,
https://doi.org/10.1007/978-3-030-85365-5_7

or the learners. Many studies seek better understanding and characteristics definition of these struggles in order to improve the introduced support for teachers. However, the struggle's definition is changeable according to the teacher's experience, the area where the teacher works, and the level of education of the learners [1, 2]. Mostly, these struggles are related to learners' reluctance to the learning process with a correlation to their social and cultural environment. Generally, young teachers especially who work in rural areas struggle more than the experienced teachers [3]. An obscuration of these struggles effect can be achieved by considering the socio-cultural context of the environment. Therefore, we must conduct an effective deduction from each teacher's context to introduce a suitable educational methodology to help this teacher by delivering a personalized experience in the form of a training course or a resources recommendation. In a teaching process's context, an educational resource can be defined as the tools to transmit the formulated educational materials to learner in the form of multimedia, textbook, course, activity, event, and expert person [4]. The main purposes of these resources are the learner motivation, extension of creativity, as well as the acquirement of additional skills. These educational resources can be used for the teaching and the learning processes. We can classify these resources according to its educational usage into instructional, collaboration, practical, and evaluation resources [5]. Nowadays, we notice the vast quantity of courses and resources which can bring the needed aid for the teacher, is not well-organized. Some resources are considered as intellectual property for the educational institution and others are open-source materials which can be used by anyone without any obligations [5]. In order to arrive to the desired results, multiple questions should be addressed; why do we select the educational resources as a solution? and how do we adapt these resources to benefit each teacher separately?

2 Problem Definition

In general, the teachers' problems are defined according to the following three factors; learner's lack of motivation, teacher's professional-personal life balance, and teacher's accurate self-evaluation [1]. The lack of motivation problem affects learners everywhere despite the learner's social conditions which makes the learner's relationship with the educational institution critical. This problem leads the teacher eventually to put more effort and energy keeping the learner's interest at high level. Such effort exerted by teachers creates some sort of work addiction which leads to inseparable professional and personal lives. In order to find a professional-personal balance, the teacher should gain the self-evaluation skills through a proper training resources to introduce efficient methodologies.

Moreover, young teachers who are working in rural areas, face three additional struggles; learner's discipline and behavior in the educational institution, working area's culture and traditions, and learner's ease of mobility in the rural areas [3]. One of the essential factors of the educational process success in rural areas is maintaining the ambitions for knowledge which is affected by the absence of discipline. As a result, young teachers often complain about learners losing their ambition because of the time wasted on keeping the order and discipline in classes. Another difficulty that impacts the learning progress in the rural areas is the learner's mobility to educational institutions. There are two reasons for such problem; one is a consequence to the learners losing motivation and interest to continue their education, and the other reason is their families' mobility from place to place. Culture is the last of the precedent problems and it means culture for both the learners and teachers. From the learner's perspective, they have hard time keeping track of the large amount of new information that is provided in the educational institution [6]. On the other hand, teachers most probably do not have prior knowledge about the culture of their working areas.

All these struggles have correspondences between each other which leads to suggesting cumulative solutions that address multiple problems at once. Figure 1 shows the concluded connection between the teacher's problems with a suggestion of cumulative solutions that address each connection. Briefly, a young teacher needs proper recommendations for suitable educational resources to strengthen course content, and appropriate training courses in order to achieve significant results in his job during the very early stage. Thus, socially, our concern is to find a solution that helps the teachers with different backgrounds and profiles.

In order to find a solution, scientifically, we found that the existence of an educational recommender system (ERS) will grant the ability to be accessible all times for teachers, and to adapt to the teacher's preferences and context of the search operation. The system adaptation to teacher's preferences can be accomplished by both monitoring teachers' activities and collecting the data which represent unique features,

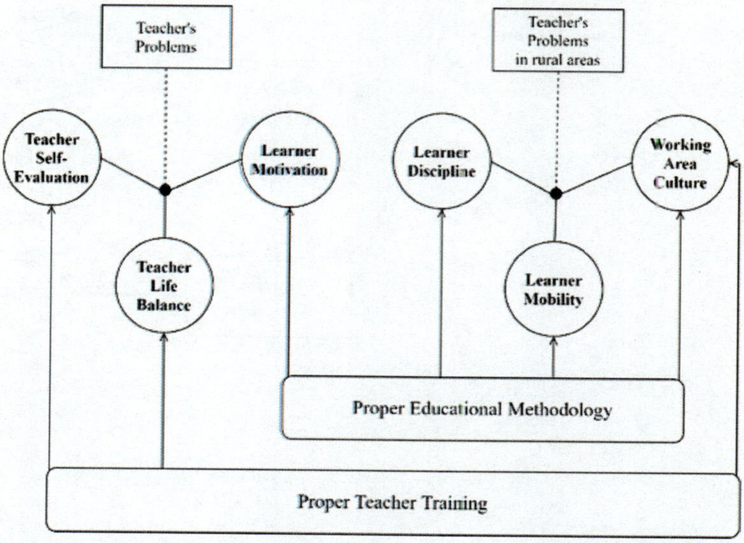

Fig. 1 Relation between teacher's struggles

and the teacher input details. This process helps defining the context of the teacher's search processes. Accordingly, in order to achieve the desired level of personalization, the system should consider number of features to define as shown in Fig. 2; the working environment and culture, the teacher's experience, and the targeted students background. The working environment can be an urban or rural area with different cultural backgrounds and traditions while the teacher's experience represents the educational level and the number of years in the teaching career. Another factor, student's background, can be either an educational level or academic degree. The last factor is these collected data from details provided by teachers, their activities while interacting with the system, and the resources characteristics [7] creating a multi-criteria problem [8].

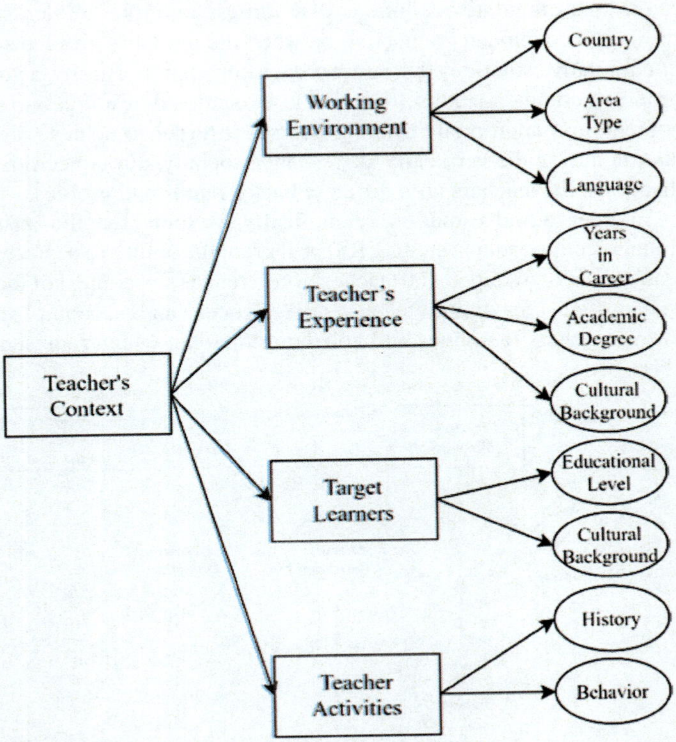

Fig. 2 Teacher's context aspects

3 Related Work

The review papers that were published during the past five years [9–11], discussed the educational recommendation systems trends from 2010 until 2018 with only 3 papers, 1 paper, and 1 paper respectively that take the teacher as the targeted user and none of these RSs took into consideration all the mentioned factors. We surveyed educational resources recommender systems that provide the teacher with either educational resources or training courses during the last ten years. After an excessive search, 11 only out of 34 ERSs met the search criteria as shown in Table 1, while the omitted ones either do not provide the precedent two types of recommendations or target learners as well as teachers. In order to conclude the essential criteria for constructing an efficient RS, four critical points should be discussed; user personalized experience, RS model, data sources, and items measurability as shown in Fig. 3 [12]. Every teacher gets resources recommendations according to the teacher profile and activities creating a realistic experience for the teacher [13]. From the surveyed papers, 4 ERSs only provide a personalized experience for teachers [7, 14–16]. As our problem has multiple aspects, therefore it is not sufficient to use one RS strategy to address all

Table 1 Summary of reviewed ERRSs

Paper	Year	Item	Data model	Approach
Manouselis et al. [20]	2010	Evaluation of current resources	European Schoolnet portal data model[a]	Collaborative-based
Bozo et al. [18]	2010	Learning Objects (LOs)	IEEE learning object metadata (IEEE LOM) [21]	Hybrid
Zhao et al. [14]	2014	Teaching Resources	Knowledge tree	Knowledge-based
Sergis et al. [7]	2014	LOs	IEEE LOM	Competence-based (hybrid)
Rodriguez et al. [15]	2015	Educational Resources	Ontology	Knowledge-based
Sabourin et al. [16]	2015	Educational Resources (SAS)	Descriptive Models (SAS Enterprise Miner)	Hybrid
Sergis and Sampson [13]	2015	Learning Objects Repositories (LORs)	–	Competence-based (hybrid)
Pursel et al. [19]	2016	Open Educational Resources (OER)	– -	Design-based (hybrid)
Villalba et al. [22]	2017	LOs	IEEE LOM	Collaborative-based
Ali et al. [23]	2017	Educational Resources	MEMORAe SoIS [24]	Hybrid
Iniesto and Rodrigo [25]	2019	Modifications to MOOC courses	Massive Open Online Courses (MOOCs) [26]	Collaborative-based

[a] http://www.eun.org/

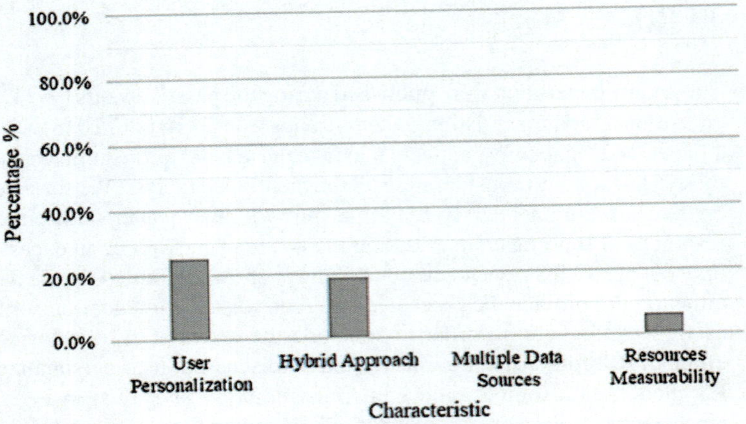

Fig. 3 Percentage of recommender systems fulfilling the required characteristics

these aspects. A hybridization of two strategies is a necessity to achieve acceptable results [17] which is followed by 3 ERSs from our list [13, 18, 19] but none of them considered the geographical dimension of the problem. All 11 ERSs use only one source of educational resources datasets which limits the availability and variety of the resources. The variety of data sources creates a reliable experience for users [20]. The last point to be discussed is the measurability of educational resources which is important to deliver the right resources to the targeted teacher. This important criterion had been taken into consideration by only one ERSs [15].

4 Our Approach

The educational resources recommender system is targeting teachers in different fields to provide educational resources that match the search process context along with the teacher profile. In order to achieve this objective, a semantic representation of the system context is needed in the form of a well-constructed ontology with the teacher context as its scope. This ontology models the teacher profile, the learner profile, the resource, and most importantly, the working and living environments in terms of geographical information and cultural nature. The system also stores the teacher's activities such as resources' view, search history, and interacting behavior into the ontology. The ontology facilitates the recommendation process of the system through matching the working environment with the living environment of both the teacher and the learner in addition to matching the teacher profile with the learners' profile. The resources database is constructed from the merging of multiple sources such as open-source educational platforms that provides different types of educational resources, global datasets that are available online, courses and materials provided by universities, along with the individual experiences shared by other teachers.

In order to conduct efficient recommendations, two approaches should be combined to construct a hybrid RS capable of introducing a dynamic experience for teachers. First, a knowledge-based approach is required to recommend resources according to the pre-configured selection factors for each type of resources in a Multi-Criteria Decision Making (MCDM) problem. The usage of MCDM is raised due to multi-factor measurement of each type of educational resources along with the teacher's profile. The MCDM searches the semantic dataset by evaluating the multiple selection criteria in order to find the best match for the user input. The multiple factors that support the resources recommendation, are supported languages by the resource, learners' age category, usage cost, rating and popularity, technical expertise level of teacher, education level of teacher, geographical proximity and location, and the existence of any additional supporting tools. Each of these factors has a different weight according to the type of the resource while calculating the relevance score as in Eq. (1) where u_i is the utility function for factor i and w_i is the corresponding weight. The factors' weights are obtained from the provided experts rating and are changeable according to the teacher personalized experience from the system usage.

$$\sum_{i=0}^{n} w_i \cdot u_i \tag{1}$$

Then, a content-based approach is triggered by the teacher's interactions with provided resources from the previous approach. This approach should use an assortment of machine learning techniques which is presented by either Neural Networks (NN) or Deep Neural Networks (DNN) to decide the next personalized recommendation based on the teacher's selection of resources. NN and DNN have the capability to predict resource factors' importance for a certain teacher from the teacher's activity log by discovering the pattern of selection for a certain teacher. The obtained results must be updated continuously according to the two approaches outputs which are elicited and rearranged. Finally, all teacher's interactions must be stored as the user feedback for future system usage. In the following example, a scenario of the system usage by a teacher is presented.

Scenario: A user with role teacher is searching for a resource:

1. The teacher enters search keyword/s for a certain resource type or select all types.
2. A query is constructed by merging the keywords with the teacher's profile.
3. The query is then processed, and results are retrieved and ordered according to the factors' weights corresponding to each resource.
4. A list of resources is displayed to the teacher.
5. The teacher selects or bookmarks preferred resource/s (user rating).
6. Resources selection array is stored into database for user profiling.

7. Then the selection array is processed to calculate how close were the teacher selections matched the expert's factors. For example,
    a. A teacher selects a resource of type event. The event, in general, has the distance factor with the highest weight.
    b. The system found out that the selected event is very far from the teacher's location.
    c. The difference between both locations are calculated and compared against the previous acceptable average distance between a teacher's and an event's locations.
    d. According to the comparison, a new factor's weight (increment or decrement) is applied and stored to the teacher's preferences. (personalized factor's weights)
8. A second round of recommendations is generated after the previous mechanism.
9. New results are retrieved according the alterations in factors' weights and in parallel the content-based approach is running to find similar resources to the selected/bookmarked ones.
10. The result page is updated dynamically with the new recommended resources (Fig. 4).

5 Conclusion and Outlook

The objective of this paper is to present a proper educational recommender system approach which can generate realistic recommendations for teachers despite their level of education, the working environment or the targeted students. First, we identify the main struggles of teachers in the surrounding environment and which of these struggles obstructs the teacher's career the most. After identifying the struggles, we concluded the suitable solutions for the teacher by recommending the proper educational methodologies and training courses for teachers. Finally, we proposed the main characteristics for an educational resources recommender system which provides the teacher with the proper solution.

After reviewing all the ERSs related to the proposed system, we conclude that helping the teacher to overcome the facing struggles in the teaching career can be achieved by a system which meets all the discussed criteria. This recommender system must include the following points;

- Merging multiple data sources into a semantic dataset
- Introducing a teacher profiling according to the targeted students.
- Predefined weighted factors for accurate resources measurability
- A hybrid combination of recommendation approaches that consider both the teacher-resource matching and the teacher's profile.

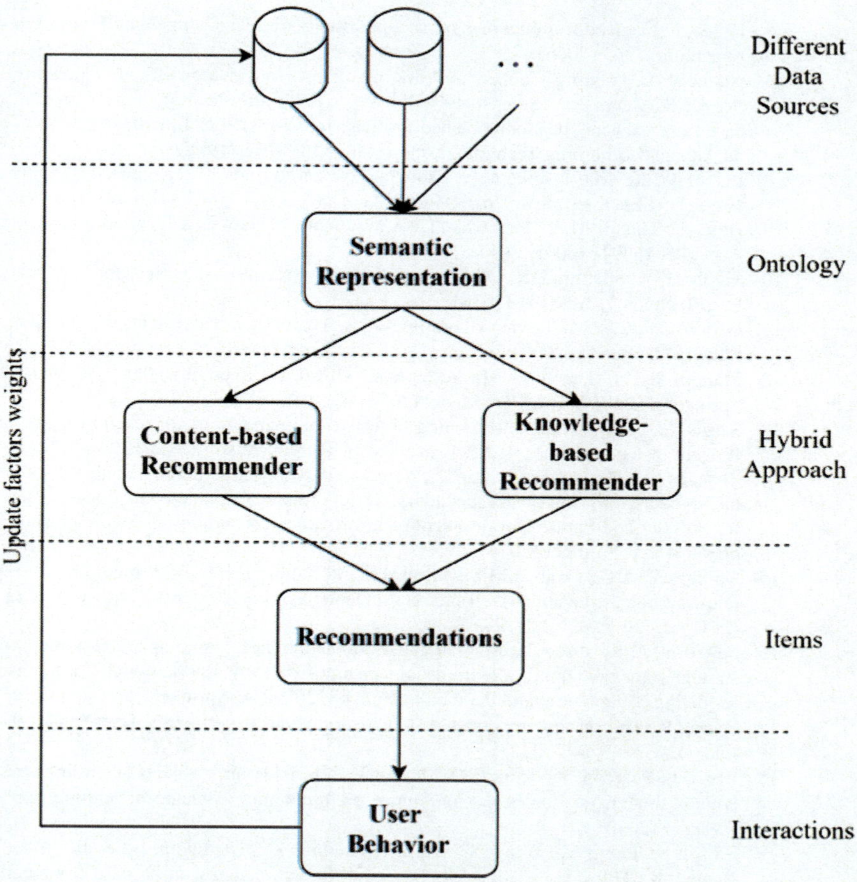

Fig. 4 Block diagram of the proposed approach

In the moment, a system which takes the following points into consideration undergoes the implementation and testing processes.

References

1. Natale, J.A.: Why teachers leave. Execut. Educat. **15**, 14–18 (1993)
2. Veenman, S.: Perceived problems of beginning teachers. Rev. Educ. Res. **54**, 143–178 (1984)
3. Monk, D.H.: Recruiting and retaining high-quality teachers in rural areas. In: The Future of Children, 155–174 (2007)
4. Bušljeta, R.: Effective use of teaching and learning resources. Czech-Polish Histor. Pedagog. J. **5**,(2013). https://doi.org/10.2478/cphpj-2013-0014

5. Hylén, J.: Open educational resources: Opportunities and challenges. In: Proceedings of Open Education, 4963 (2006)
6. Auduc, J.-L. Training Teachers to Work in Schools Considered Difficult. Citeseer (1998)
7. Sergis, S., Zervas, P., Sampson, D.G.:x (2014) Towards learning object recommendations based on teachers' ICT competence profiles. In: 2014 IEEE 14th International Conference on Advanced Learning Technologies, pp. 534–538. IEEE (1998)
8. Lakiotaki, K., Matsatsinis, N.F., Tsoukias, A.: Multicriteria user modeling in recommender systems. IEEE Intell. Syst. **26**, 64–76 (2011)
9. Dutt, A., Ismail, M.A., Herawan, T.: A systematic review on educational data mining. Ieee Access **5**, 15991–16005 (2017)
10. Vaidhehi, V., Suchithra, R.: A systematic review of recommender systems in education. Int. J. Eng. Technol. **7**, 188–191 (2018)
11. George G, Lal, A.M.: Review of ontology-based recommender systems in e-learning. Comput. Educ. **142**, 103642 (2019)
12. Manouselis, N., Drachsler, H., Verbert, K., Duval, E.: Recommender Systems for Learning. Springer Science & Business Media (2012)
13. Sergis, S., Sampson, D.G.: Learning object recommendations for teachers based on elicited ICT competence profiles. IEEE Trans. Learn. Technol. **9**, 67–80 (2015)
14. Zhao, M., Ni, W., Zhang, H., Lin, Z., Yang, Y.: A Knowledge-based teaching resources recommend model for primary and secondary school oriented distance-education teaching platform. In: Frontier and Future Development of Information Technology in Medicine and Education, pp. 511–521. Springer (2014)
15. Rodrıguez, A.C., Gago, J.M.S., Rifón, L.E.A., Rodrıguez, R.P.: A recommender system for non-traditional educational resources: a semantic approach. J. Univ. Comput. Sci. **21**, 306–325 (2015)
16. Sabourin, J., Kosturko, L., McQuiggan, S.: Where to next? A comparison of recommendation strategies for navigating a learning object repository. In: International Conference on User Modeling, Adaptation, and Personalization, pp. 208–215. Springer (2015)
17. Burke, R.: Hybrid recommender systems: survey and experiments. User Model. User-Adap. Inter. **12**, 331–370 (2002)
18. Bozo, J., Alarcón, R., Iribarra, S.: Recommending learning objects according to a teachers' context model. In: European Conference on Technology Enhanced Learning, pp. 470–475. Springer (2010)
19. Pursel, B., Liang, C., Wang, S., Wu, Z., Williams, K., Brautigam, B., Saul, S., Williams, H., Bowen, K., Giles, C.L.: BBookX: Design of an automated web-based recommender system for the creation of open learning content. In: Proceedings of the 25th International Conference Companion on World Wide Web, pp. 929–933 (2016)
20. Manouselis, N., Vuorikari, R., Van Assche, F.: Collaborative recommendation of e-learning resources: an experimental investigation. J. Comput. Assist. Learn. **26**, 227–242 (2010)
21. RISK, U. Draft standard for Learning Object Metadata. IEEE Standard 1484 (2002):
22. Villalba, K., Cuba, S.C., Deco, C., Bender, C., García-Peñalvo, F.J.: A recommender system of open educational resources based on the purpose of learning. In: 2017 Twelfth Latin American Conference on Learning Technologies (LACLO), pp. 1–4. IEEE (2017)
23. Ali, B., Majd, S., Marie-Hélène, A., Elsa, N. Recommendation of pedagogical resources within a learning ecosystem. In: Proceedings of the 9th International Conference on Management of Digital EcoSystems, pp. 14–21 (2017)
24. Abel, M.-H., Benayache, A., Lenne, D., Moulin, C.: E-MEMORAe: a content-oriented environment for e-learning. In: E-Learning Networked Environments and Architectures, pp. 186–205. Springer (2007)
25. Iniesto, F., Rodrigo, C.: YourMOOC4all: a recommender system for MOOCs based on collaborative filtering implementing UDL. In: European Conference on Technology Enhanced Learning, pp. 746–750. Springer (2019)
26. Porter, S.: To MOOC or Not to MOOC: how Can Online Learning Help to Build the Future of Higher Education? Chandos Publishing (2015)

Hybrid Cloud Native Framework for Smart City Applications

Amit Kar

Abstract The smart city authorities usually host the IoT applications in on-premise datacentres or partially on cloud depending upon type of applications, rules, regulations and safety of data. Smart city services are offered free to citizen, since majority of the services are intended for citizen. Managing the numerous IoT applications in on-premise data centre brings huge cost along with difficulties in application management. It is very much essential to bring some of the applications onto cloud to optimise the cost and ease the operations. The objective of the present paper is to build sustainable smart city through cost effective cloud native solution. The paper intends to analyse the cloud native approach of the smart city solution with respect to traditional smart city architecture and also suggests a hybrid approach for cost effective future smart city solutions.

Keywords IoT · Cloud · Datacenters

1 Introduction

Cities are being transformed into smart cities by providing smart solutions to make citizens' lives better. Transforming a city into a smart city needs the right vision to improve livelihood translating through different smart solutions and services. The objective of the smart city is to influence economic growth and transform the lives of the citizen by providing smart solutions and services.

Worldwide smart city solutions based on the Internet of Things (IoT) are transforming citizen day to day lives. We live in an exceptionally networked world. In the current time rapid growth of the use of Internet of Things (IoT) devices makes it even more complex. A sustainable smart city is one that continuously adopts new innovative models and augments its solutions to serve citizen in an efficient way.

With the current COVID pandemic, people have realised the importance of technology and how it is important to adapt the cloud technologies, AI/ML, IoT and

A. Kar (✉)
Sterlite Technologies Limited, Gurugram, India
e-mail: amit.kar@stl.tech

L. Troiano et al. (eds.), *Advances in Deep Learning, Artificial Intelligence and Robotics*, Lecture Notes in Networks and Systems 249,
https://doi.org/10.1007/978-3-030-85365-5_8

other evolving technologies for efficiency, flexibility and effectiveness. Cloud adoption has picked up the pace and has shown the way to deal with difficult situations of increasing infrastructure, faster deployment, resiliency and operational expenses.

Businesses are actively watching their Capex and Opex cost with fluctuations in demand and they are willing to host their applications on the cloud that were earlier on their on-premise self-managed infrastructure. With public cloud providers (AWS/Azure/GCP) pay-as-you-go models for multiple services like IaaS, PaaS and SaaS, it is easier to provision, maintain and deploy applications at scale and creating an efficient infrastructure.

Cloud platforms have matured with respect to software-defined networks, security, virtualization, containers, automatic back-up, resiliency and have given the confidence to move organizations on fully public or hybrid (public + private) kind of environment.

Smart city applications have different usage and usability scenarios and there is no all-in-one fit model. Applications taking a lot of bandwidth and low latency will have to be closer to the source. Take an example of computer vision applications (Facial recognition, Object detection, loitering, crowd detection etc.) which require video streams to be worked upon and need instant processing.

Similarly, there are IoT applications that increase gradually with increasing customers/citizens. Smart water meters and smart electricity meters are some of the applications which increase with increase of consumer base. These applications can also be deployed on public cloud infrastructure which gives better agility and efficiency of resources.

The smart city authority usually hosts the IoT applications in on-premise datacentres or partially on the cloud depending upon the type of applications, rules and regulation and safety of data. Smart city services are offered free to the citizen, since the majority of the services are intended for the citizens. Managing the numerous IoT applications on-premise data centre brings huge costs along with difficulties in application management. It is very much essential to bring some of the applications onto the cloud to optimise the cost and ease the operations.

The objective of the present paper is to build a sustainable smart city through cost effective cloud native solution. The paper intends to analyse the cloud native approach of smart city solutions with respect to traditional smart city architecture and suggests a hybrid approach for the future smart city. In the present study, cost estimation for Smart Energy Meter System, Smart Transport, Ticketing & Parking System and mart Garbage Detection & Collection System has been done using both clod native and data center technique.

2 Methodology

Generally, the complete smart city solution is deployed in multiple layers such as IoTs/Sensors layer, network layer and applications/IoT platform layer (Fig. 1). The

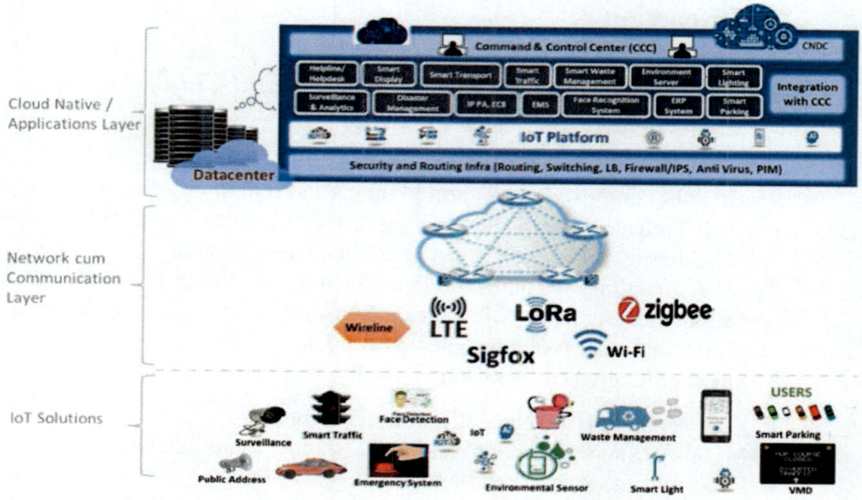

Fig. 1 Traditional smart city architecture

sensors are connected to the applications layer through a network layer. These are being monitored and managed by the hosted in the on-premise datacentre at central.

The Services or the applications like Smart Parking, Smart Lighting, Smart Waste Management System, Automatic Traffic Control systems, Smart Health system, Surveillance, smart energy and Safety System are being offered as part of smart city solutions.

In the majority of the cases, these applications are monolithic in nature and hosted in the on-premises datacentre as mentioned in the above Fig. 1. In some cases some of these monolithic applications are hosted in the pubic cloud but it consumes a lot of infrastructures. Hence, it incurs a lot of cost to smart city authorities.

In this paper, a futuristic approach is suggested with a Hybrid model combining both legacy applications in the on-premise Datacentre and micro service based applications on the cloud to get the maximum benefit.

In this case study, the Legacy monolithic applications are containerised to achieve scalability and micro service-based applications are shifted to the cloud. By doing this the whole solution gets the benefit of a complete cloud native approach. Effective Cloud Native applications are identified in terms of flexibility in operation, cost competitiveness, scalability.

Case studies comprising of Smart Energy Meter System, Smart Transport, Ticketing & Parking System and mart Garbage Detection & Collection System have considered estimating the cost competitiveness.

3 Discussion

3.1 Connection Between Smart Cities and Cloud Native Application

Smart cities comprise interconnected things and multiple applications ranging from Smart IoT elements of Smart Parking, Smart Metering, Smart Lighting to Smart policing through cameras. The list of applications that smart city imbibes is exhaustive and sometimes becomes difficult to maintain. Authority managing smart city operations face challenges with cities growing infrastructure & population generating multiple events, alerts and insights to act upon.

Smart cities applications can be broadly categorized into the following:

- E-Governance (Citizen applications, G2C portals)
- Mobility (Smart transport, Automatic Number Plate Recognition)
- Environment (Smart Environment sensors, Pollution sensors)
- Security (Facial recognition, Video analytics, Intrusion detection, Object detection)
- Living (Smart water, Smart Metering, Waste detection, Water and Sewage leakage)

Emerging technologies and the cloud native applications are perfect driving force for smart cities applications, which need scalable, open, interconnected systems for offering near real-time services and analytics on big data. Some of the areas in which cloud computing can help smart cities:

- Agility (Deploying application changes faster)
- Faster deployment (Micro-services based applications, CI/CD—e.g. Covid app)
- PaaS (Platform as a service—containers, databases etc.)
- Open platforms (API-based e.g. G2C services)
- Security and compliance (as per demands of local by-laws)
- Lower TCO (On-demand provisioning and scalability e.g. Increasing demand of security applications server infra during events like large social gathering, festive occasion, rallies etc.)

3.2 Hybrid Cloud Native Architecture

While evaluating the approach of cloud native for smart city solutions it's found that some of the applications by design and by nature are favouring to cloud native.

In the deployment of large smart city applications, many services and IoT applications will need to be deployed based on a number of online requests by citizens/city organizations. Microservices running on the cloud platform will give flexibility for easy and formalized access to applications. An underlying elastic infrastructure which is scaled horizontally and vertically to cater to the increasing demands will provide a platform for containerized IoT applications. Some of the smart solutions which are consumer driven such as smart parking, smart challan system, and Facial recognition system need instant scaling up and scaling down of underlay hardware infrastructure. Modern applications like Video Analytics (VA)/ Video management Systems (VMS) are constantly evolving. These applications have moved to micro-services and containers for efficient use of resources and scalability.

As an example, if there is an event happening in a city where Video Analytics systems need to analyse faces from the crowd, then the situation demands an instant load on the Facial Recognition System (FRS) and instant scalability is needed. Micro services will enable systems to scale individually by not affecting other VA applications (Object detection, Loitering etc.). Investment in permanent hardware for probable demand in on-premise datacentre or on cloud leads to huge costs. A model architecture of Cloud native smart solutions is depicted in Fig. 2.

By shifting these applications to the cloud with micro service based architectures gets the benefit of cloud such as cost optimisation due to pay as-use model, Scalability as on when demand comes, Flexibility to develop and update any modules of the applications.

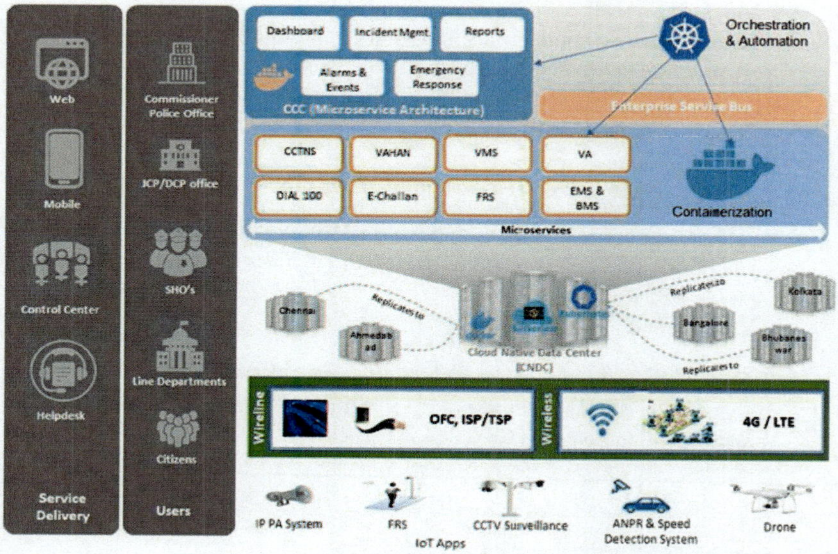

Fig. 2 Hybrid cloud native architecture of smart city solution

3.3 Cost Estimation: A Case Study

Applications that are cloud native and are based on microservices architecture are changing the way applications are deployed, scaled and replicated. These architectural patterns can be effective in providing an application stack for smart cities that can be replicated in multiple similar cities and learnings from one can be transferred to another.

Security is one of the key requirements of a smart city and it includes multiple applications such as facial recognition, object detection, crowd detection, loitering and many other women safety applications. Smart city government organizations want the flexibility to run any of these applications on any camera as per their requirement to monitor a particular area/event. Every individual application doesn't have to be running at its full potential and needs to be scaled on demand which can be done easily within a microservices architecture. This helps in reducing capital cost and operating cost of infrastructure related to information and communication technologies.

The estimated cost for Garbage Detection & Collection System, Smart Transport, Ticketing & Parking System and Smart Energy Meter System are given in Tables 1, 2 and 3 respectively.

Citizen centric applications like smart water/energy meter applications for which infrastructure has to be procured gradually over years can save nearly $ 0.5 Million in a period of 5 years by adopting a cloud deployment model rather than traditional on-prem data centres. (**Assumptions**: Smart meter installation from 1 to 10L gradually y-o-y).

Other similar citizen centric applications and tentative saving:

- Smart waste Management application: Reduction of around $ 0.15 Million over a period of 5 years (Assumption: *100 to 1.2 K sensors added gradually over years with 50% utilisation*)

Table 1 Cost estimation for Smart garbage detection & collection system (800 bins & 400 vehicles)

Year of project operation	Compute (in cores)	RAM (in GB)	NVMe SSD (in GB)	Public cloud (in USD)	On-prem DC (in USD)	Cost difference (in USD)
1st Year	16	32	400	$ 3479	$ 72,055	**$ 68,576**
2nd Year	32	64	800	$ 6957	$ 21,781	**$ 14,824**
3rd Year	64	128	1600	$ 13,914	$ 21,781	**$ 7867**
4th Year	96	192	2400	$ 20,871	$ 21,781	**$ 909**
5th Year	128	256	3200	$ 27,828	$ 21,781	**$ −6048**
Cost savings (in a period of 5 years)						**$ 86,128**

Note The above public cloud cost has taken from Amazon cloud estimation calculator https://calculator.s3.amazonaws.com/index.html (Linux on c5d.4xlarge with 50% utilisation). It may vary from one cloud providers to other

Table 2 Cost estimation for smart transport, ticketing & parking system (800 vehicle)

Year of project operation	Compute (in cores)	RAM (In GB)	NVMe SSD (in GB)	Public cloud (in USD)	On-prem DC (in USD)	Cost difference (in USD)
1st Year	16	64	600	$ 3306	$ 1,08,205	**$ 104,900**
2nd Year	32	128	1200	$ 6612	$ 29,219	**$ 22,607**
3rd Year	64	256	2400	$ 13224	$ 29,219	**$ 15,996**
4th Year	96	384	3600	$ 19,835	$ 29,219	**$ 9,384**
5th Year	128	512	4800	$ 26,447	$ 29,219	**$ 2772**
Cost savings (in a period of 5 years)						**$ 155,659**

Note The above public cloud cost has taken from Amazon cloud estimation calculator https://calculator.s3.amazonaws.com/index.html (Linux on m5ad.4xlarge with 70% utilisation). It may vary from one cloud providers to other

Table 3 Cost estimation for smart energy meter system (10L smart meters)

Year of project operation	Meter connection to households (lakh meters)	Compute (in cores)	RAM (in GB)	NVMe SSD (in GB)	Public cloud (in USD)	On-prem DC (in USD)	Cost difference (in USD)
1st Year	1	48	384	1800	$ 17,507	$ 456,205	**$ 438,699**
2nd Year	2.5	96	768	3600	$ 35,013	$ 100,877	**$ 65,864**
3rd Year	5	192	1536	7200	$ 70,026	$ 100,877	**$ 30,850**
4th Year	7.5	288	2304	10,800	$ 1,05,039	$ 100,877	**$ −4,163**
5th Year	10	384	3072	14,400	$ 1,40,052	$ 100,877	**$ −39,176**
Cost savings (in a period of 5 years)							**$ 492,075**

Note The above public cloud cost has taken from Amazon cloud estimation calculator https://calculator.s3.amazonaws.com/index.html (Linux on r5ad.12xlargem with 100% utilisation). It may vary from one cloud providers to other

- Smart transport and ticketing system: TCO reduction of $ 0.1 Million over a period of 5 years (Assumption: *100 to 800 sensors added gradually over years with 70% utilisation*)

From the current case study, it is seen that smart energy meter application is being expanded on instant demand basis due to digitalisation. Hence, instant underlying hardware is to be scaled, which is not feasible on premise data center. In cloud additional hardware can be assigned instantly. On the basis of cost component, for a long time period, underlay hardware required for smart energy meter application may not be beneficial.

Multiple interconnected systems in smart cities require a central dashboard which provides insights and actionable insights from the data received at the command and control centre. Integrated command and control centre (ICCC) is the heart of smart cities operations and infrastructure. Moving systems to a unified private/hybrid cloud architecture will provide a centralised shareable data repository and will enable multiple cities to consume and share data within themselves.

Unified cloud will enable standardisation in smart city applications with on-demand services enabled for cities to consume which have limited budgets. Standardisation will also enable city services, applications for an easy infusion of new technologies and uptake of proven smart city solutions.

4 Conclusion

There is a necessity to create smart cities architecture more flexible that can cope up with different environments, increasing number of devices, users and services. In order to manage a variety of devices from different vendors, managing multiple protocols with varied applications, microservices architecture has been increasingly adopted in many IoT based solutions.

Microservices architecture leads to cloud native approach which involves loosely coupled services, scalable components which enables easy maintenance and rollout of services. Microservices bundled in containers and deployed on the cloud provides an extra edge in easy rollouts and replicating a same environment at multiple data centre/locations with a scale on demand capabilities. Modern applications are adopting microservices and container based deployment but there are legacy applications that can't be moved to microservices instantly. These legacy applications can be containerized (Fast-monolithic applications) to achieve faster deployment and consistent experience.

The hybrid approach will not only help to make a smart city sustainable but also bring quick benefits to citizens and all stakeholders. Cost, scalability and agility in deployment play a major role in the selection of hybrid architecture.

A Two-Stage Intrusion Detection System (TIDS) for Internet of Things

Monika Vishwakarma and Nishtha Kesswani

Abstract With the emphasis on data-driven decisions, the use of Internet of Things (IoT) devices to collect data has increased significantly. Because of the small size, IoT devices can be easily deployed in remote areas where continuous monitoring is not practically possible. To detect these, we have proposed a two-stage intrusion detection technique to identify attacks on the network, such as local access from remote and user to root access. Our proposed model uses the Naive Bayes classifier as the first step of intrusion detection and then passes the records that pretend to be normal for the second stage of the classifier, i.e., the k-means. We have shown results based on NSL-KDD standard data set. The results indicate an accuracy of 86.46% of the proposed approach.

Keywords Intrusion detection system · IoT · Two stage classification · R2L · U2R

1 Introduction

INTERNET has significantly impacted the lifestyle in today's era. A major revolutionary technology enabled by the Internet, the Internet of Things [9] has the capable of connecting billions of devices and systems to the Internet. However, Current IoT networks are vulnerable to cyber attacks due to the diversity of smart devices connected to the web [20]. Any intrusion [13] on the network or on the architecture layers of the IoT may lead to the serious issues, and violate the security policies like, Confidentiality, Integrity and Availability of the data.

An Intrusion Detection System (IDS) actively monitors the network for malicious activities and alerts when it detects attacks. According to the most prominent datasets used for IDS: KDD, KDDCUP99, and NSL-KDD dataset [7] attacks are separated

M. Vishwakarma (✉) · N. Kesswani
Central University of Rajasthan, Ajmer, India

L. Troiano et al. (eds.), *Advances in Deep Learning, Artificial Intelligence and Robotics*, Lecture Notes in Networks and Systems 249,
https://doi.org/10.1007/978-3-030-85365-5_9

into four main categories based on network rooted on intrusion. The attack categories are as follows:

- **Probe**: It is an attack in which the attacker scans a networking device in order to determine vulnerabilities.
- **Denial of Service (DoS)**: It is an intentional attempt to disrupt the normal traffic of the network and tries to make the network resource busy or makes it inaccessible to its intended users.
- **User to Root (U2R)**: Obtaining access to superuser rights of the local machine without legit permissions.
- **Root to Local (R2L)**: Gaining access from remote machines without legit permissions.

The main challenging task is to identify the U2R and R2L attacks in the pre-processed NSL-KDD dataset. Because the test data has new attacks that are unknown for the training dataset and due to imbalanced nature of dataset, U2R and R2L attacks start behaving as normal on the network, which is very harmful for the IDS. Many researchers have achieved a reasonable accuracy rate by applying various machine learning algorithms, but attacks like U2R and R2L have been ignored. So we propose a two-stage intrusion detection system that mainly targeted U2R and R2L attacks.

In this paper, we used machine learning algorithms so as to achieve better results with the low computation complexity which is an essential factor for the resource constrained IoT devices.

The rest fabrication of this paper is prepared as follows. Section 2 describes the Related work. The proposed two-stage intrusion detection system (TIDS) is discussed in Sect. 3. The experimental setup and results are presented in Sect. 4. The results demonstrate that the performance has increased for our proposed TIDS model as compared to other existing techniques.

2 Related Work

Several researchers have tried to suggest Intrusion Detection systems for IoT devices. Pajouh et al. [17] proposed a model in which they perform dimension reduction and classification in Two-layer called PCA, LDA, and Naive Bayes, CF-KNN, respectively. They identified four types of attacks, such as DoS, Probe, and mainly targeted U2R and R2L.

To reduce malicious activity, Moustafa et al. [14] proposed an ensemble IDS based on established statistical flow features, especially botnet attacks against MQTT, DNS, and HTTP protocols used in IoT networks.

Li et al. [11] proposed feature extraction, including a deep learning model with intrusion detection technology for smart cities. The experiments were conducted on the KDD CUP 99 dataset.

Aljawarneh et al. [3] proposed a hybrid based model. First, they filtered the data using a voting algorithm with information gain to select the crucial characteristics.

After that, they used J48, Meta Pagging, Random Tree, REPTree, AdaBoostM1, DecisionStump, Naíve Bayes classification algorithms to classify the anomaly.

Anthi et al. [4] proposed a three-layer IDS for cyber-attack detection on the IoT network. The effect of architecture was estimated by extending the four primary attacks to 12 attacks. These four attacks are DoS, Man-In-The-Middle/Spoofing, Reply, and Reconnaissance.

Li et al. [12] presented an AI-based intrusion detection in two-stages, enabled by a software-defined technique. They classified the flow using random forest within an adaptive change of the weight of the samples using a weighted voting mechanism.

Chaudhary et al. [6] designed detection and prevention algorithms, i.e., key match algorithms and cluster-based approaches to detect the routing attacks, namely selective forwarding and sinkhole attacks in the IoT network. They improved the detection rate of intrusion on the same attack by proposing a cluster-based algorithm [5].

Kesswani and Agarwal [10] presented a cluster-based IDS for Smart homes. The system can detect the malicious events within the network as well as outside the network.

The majority of the intrusion detection mechanisms mentioned in the literature are unable to detect minority attacks. These include the distribution of R2L and U2R attacks as the classes of the NSL-KDD train data set are much lower than the normal and other attack classes, as in Fig. 1. It may affect the detection capability of the intrusion detection system. To address these issues, we propose a model that targets U2R and R2L attacks.

	Normal	DoS	Probe	U2R	R2L
Train	67343	45927	11656	52	995
Test	9711	7460	2421	67	2885

Fig. 1 NSL-KDD dataset class distribution

3 Two Stage Intrusion Detection System (TIDS)

The proposed two-stage IDS model consists of the two-stages of dimensionality reduction and two-stages of classification techniques, as shown in Fig. 2. It uses 1 for TCP, 2 for UDP and, 3 for ICMP, a unique value for each nominal feature in data preprocessing. The value of each feature is standardized by using Eq. 1.

$$z = (x - \boldsymbol{\mu})/\sigma \tag{1}$$

where x is the input value, μ and $\boldsymbol{\sigma}$ represents mean and standard deviation of the corresponding feature, respectively.

Before applying data to the classifier module, we need to transform data into lower dimension space to address high-dimensional data issues. We deployed both Principal Component Analysis (PCA) and Linear Discriminant Analysis (LDA) (i.e., a multivariate unsupervised dimension reduction technique and supervised dimension reduction technique, respectively) to select the most relevant features for the classification. PCA [1] extracts almost all relevant characteristics from the data by sorting the eigenvectors in descending order and choose the top k features. PCA transformed data is not ideal for classification, so we utilize LDA predictive modeling algorithm for multi-class classification. LDA [2] examines the class labels and transforms the data into $N_c - 1$ dimensions, where N_c is the number of main attack class category. Thus, the new mapped charactersics will have only four dimensions lda1, ..., lda4. Figure 3 shows the four-dimensions of the mapped original NSL-KDD train and test data set reconstructed by LDA.

Next, in the two stage IDS, we deployed Naive Bayes classification technique and K-means clustering technique [18] to enhance the performance of the IDS. The Naíve Bayes classifier is a simplified, strong Bayesian classification model [15]. The reason why Naive Bayes has been used is that the probability of one feature does not affect the probability of the other feature. Equation 2 calculates the probability of each class level y with the each feature value x.

$$P(x_i|y) = \frac{1}{\sqrt{2\boldsymbol{\pi}\boldsymbol{\sigma}_y^2}} e^{-\frac{(x_i-\mu_y)^2}{2\sigma_y^2}} \tag{2}$$

where 'μ' represents mean and '$\boldsymbol{\sigma}$' represents standard deviation.

After detecting the anomalies by passing data to the first stage of the classifier, we passed normal type data to the second stage to boost the efficiency that is detection rate, we have used K-means clustering technique. In the K-means model, we initially set centroid of each main attack (e.g., DoS, Probe, R2L, U2R, and Normal) using k-means++ to avoid the local optimum.

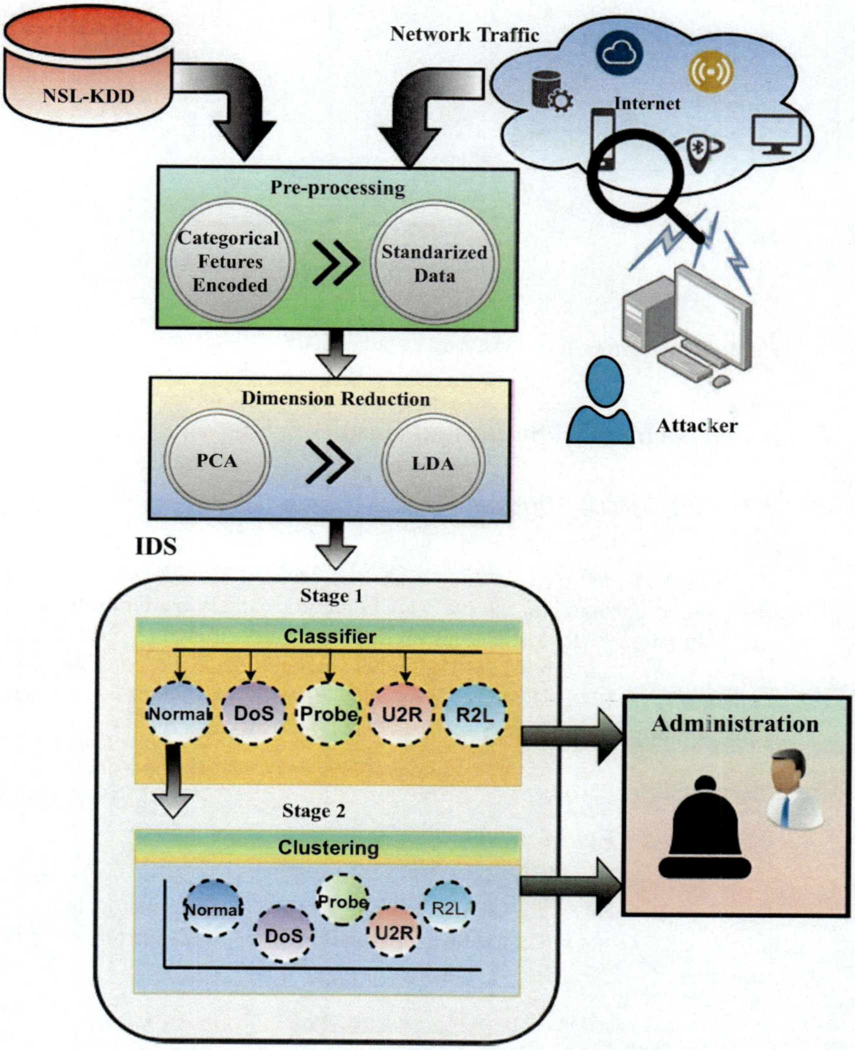

Fig. 2 The proposed IDS

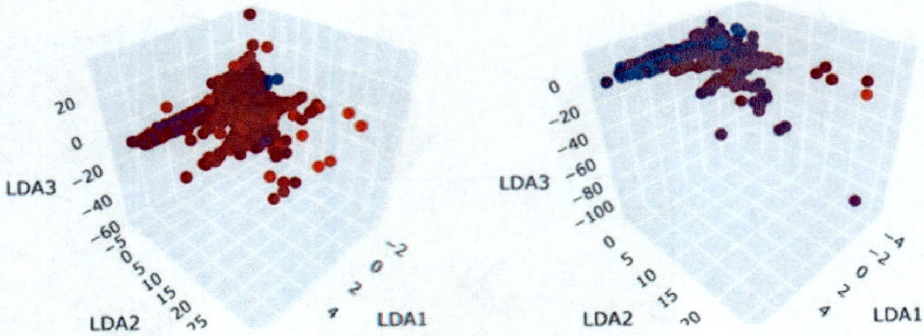

Fig. 3 LDA transform train and test data representation

4 Experimental Setup and Results

4.1 NSL-KDD Dataset

The NSL-KDD [8] data set is an improved version of the KDD-99 data set. Forty-one defined attributes in the NSL-KDD record, namely as duration, source_bytes, and destination_bytes, labeled as 39 different types of attacks. NSL-KDD data are classified into four types based on attacks, namely DoS, Probe, U2R and R2L. As shown in Table 1 NSL-KDD test data set contains 17 new attacks which are not there in the training data set.

4.2 Model Evaluation

In our experiment, we evaluate our model by calculating the accuracy, precision, recall, and f-score, which is the proper measurement to determine the model's performance.

Table 1 NSL-KDD data set attack labels in train and test data

Attacks	Sub attacks in train dataset	New sub attacks in test dataset
DoS	neptune, back, land, pod, smurf, teardrop	worm, apache2, processtable, mailbomb, udpstorm
Probe	ipsweep, satan, portsweep, nmap	saint, mscan
U2R	buffer_overflow, loadmodule, perl, rootkit	xterm, ps, sqlattack
R2L	ftp_write, spy, warezclient, warezmaster, imap, guess_passwd, multihop, phf	httptunnel, snmpguess, sendmail, named, snmpgetattack, xlock xsnoop

Accuracy indicates the correctly predicted attacks over all the data records.

$$Accuracy = \frac{T_P + T_N}{T_P + T_N + F_P + F_N} \tag{3}$$

T_P = True Positive, T_N = True Negative
F_P = False Positive, F_N = False Negative.

Precision indicates the predicted attacks which are relevant.

$$Precision = \frac{T_P}{T_P + F_P} \tag{4}$$

Recall indicates the total relevant results correctly classified by the algorithm.

$$Recall = \frac{T_P}{T_P + F_N} \tag{5}$$

F-score evaluate the model using both Precision and Recall and gives a better measure if imbalanced class distribution exists.

$$F\text{-}score = 2 \times \frac{(Precision \times Recall)}{(Precision + Recall)} \tag{6}$$

4.3 Results

In this section, we discuss the results of the experiments performed over NSL-KDD dataset. The Accuracy of the proposed IDS was compared to other existing approaches. Figure 4 shows the obtaineed precision, recall, and f-score and Table 2 shows the overall detection rate for the attack classes for the proposed IDS and other existing approaches.

In TIDS, the complexity is reduced to half by projecting 41 features into 21 features using Principal component analysis (PCA) during the first extraction process. In the second stage of feature selection, LDA converts 21 features into four features based on class labels, which is very less for the further classification process. The complexity of the Naive Bayes classification algorithm is $O(n_r \times f)$, where n_r represents the number of records in data set and f represents the number of features (i.e., only four features). In the next stage of classification, K-means is $O(n_c \times n_c)$, where n_c represents the record count in the data set, which is very less because we pass only those records that are classified as normal in the first stage of the classifier.

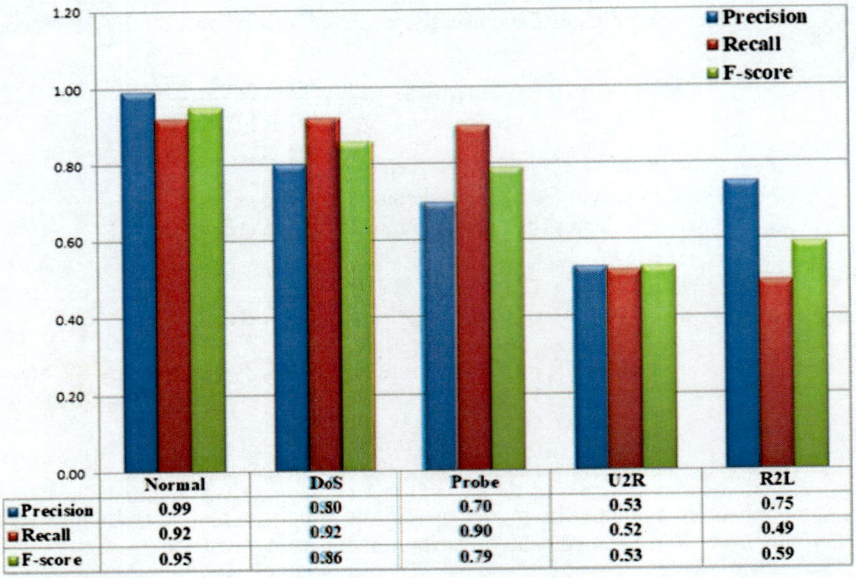

Fig. 4 Precision, recall and f-score of TIDS under different attack classes

Table 2 Accuracy of the proposed model and other existing approaches

Methods	Accuracy	Normal	DoS	Probe	U2L	R2L
Proposed TIDS	86.46	92.1	92.3	90.5	52.2	49.2
TDTC [17]	84.86	94.43	88.20	87.32	70.15	42
DBN [19]	80.58	95.64	87.96	72.97	0	0
S-NDAE [19]	85.42	97.73	94.58	95.67	2.70	2.82
Two-tier [16]	82	94.56	84.68	79.76	67.16	34.81

5 Conclusion

In this paper, we have discussed problems related to minority attacks including R2L and U2R. In order to address attacks on IoT networks that pretend themselves as a normal behavior on the network, we have come up with a proposed two-stage intrusion detection system technique. It takes the network traffic and performs the dimensionality reduction using PCA and LDA to consider only the high throughput features for further processing. Naive Bayes classification is done on the first stage and k-means clustering in the second stage to attain 86.46% accuracy.

In the future, we will improve the feature engineering technique and deploy this approach in real-time for the network traffic analysis and evaluate the performance.

References

1. Abdi, H., Williams, L.J.: Principal component analysis. Wiley Interdiscip. Rev. Comput. Stat. **2**(4), 433–459 (2010)
2. Aburomman, A.A., Reaz, M.B.I.: Ensemble of binary SVM classifiers based on PCA and LDA feature extraction for intrusion detection. In: 2016 IEEE Advanced Information Management, Communicates, Electronic and Automation Control Conference (IMCEC), pp. 636–640. IEEE (2016)
3. Aljawarneh, S., Aldwairi, M., Yassein, M.B.: Anomaly-based intrusion detection system through feature selection analysis and building hybrid efficient model. J. Comput. Sci. **25**, 152–160 (2018)
4. Anthi, E., Williams, L., Słowińska, M., Theodorakopoulos, G., Burnap, P.: A supervised intrusion detection system for smart home IoT devices. IEEE Internet Things J. **6**(5), 9042–9053 (2019)
5. Choudhary, S., Kesswani, N.: Cluster-based intrusion detection method for internet of things. In: 2019 IEEE/ACS 16th International Conference on Computer Systems and Applications (AICCSA), pp. 1–8 (2019)
6. Choudhary, S., Kesswani, N.: Detection and prevention of routing attacks in internet of things. In: 2018 17th IEEE International Conference on Trust, Security and Privacy in Computing and Communications/12th IEEE International Conference on Big Data Science and Engineering (TrustCom/BigDataSE), pp. 1537–1540. IEEE (2018)
7. Choudhary, S., Kesswani, N.: Analysis of KDD-Cup'99, NSL-KDD and UNSW-NB15 datasets using deep learning in IoT. Procedia Comput. Sci. **167**, 1561–1573 (2020)
8. Dhanabal, L., Shantharajah, S.: A study on NSL-KDD dataset for intrusion detection system based on classification algorithms. Int. J. Adv. Res. Comput. Commun. Eng. **4**(6), 446–452 (2015)
9. Evans, D.: The internet of things: how the next evolution of the internet is changing everything. CISCO White Pap. **1**(2011), 1–11 (2011)
10. Kesswani, N., Agarwal, B.: SmartGuard: an IoT-based intrusion detection system for smart homes. Int. J. Intell. Inf. Database Syst. **13**(1), 61–71 (2020)
11. Li, D., Deng, L., Lee, M., Wang, H.: IoT data feature extraction and intrusion detection system for smart cities based on deep migration learning. Int. J. Inf. Manage. **49**, 533–545 (2019)
12. Li, J., Zhao, Z., Li, R., Zhang, H.: AI-based two-stage intrusion detection for software defined IoT networks. IEEE Internet Things J. **6**(2), 2093–2102 (2018)
13. Liao, H.J., Lin, C.H.R., Lin, Y.C., Tung, K.Y.: Intrusion detection system: a comprehensive review. J. Netw. Comput. Appl. **36**(1), 16–24 (2013)
14. Moustafa, N., Turnbull, B., Choo, K.K.R.: An ensemble intrusion detection technique based on proposed statistical flow features for protecting network traffic of internet of things. IEEE Internet Things J. **6**(3), 4815–4830 (2018)
15. Mukherjee, S., Sharma, N.: Intrusion detection using Naive Bayes classifier with feature reduction. Procedia Technol. **4**, 119–128 (2012)
16. Pajouh, H.H., Dastghaibyfard, G., Hashemi, S.: Two-tier network anomaly detection model: a machine learning approach. J. Intell. Inf. Syst. **48**(1), 61–74 (2017)
17. Pajouh, H.H., Javidan, R., Khayami, R., Ali, D., Choo, K.K.R.: A two-layer dimension reduction and two-tier classification model for anomaly-based intrusion detection in IoT backbone networks. IEEE Trans. Emerg. Top. Comput. Intell. (2016)
18. Peng, K., Leung, V.C., Huang, Q.: Clustering approach based on mini batch kmeans for intrusion detection system over big data. IEEE Access **6**, 11897–11906 (2018)
19. Shone, N., Ngoc, T.N., Phai, V.D., Shi, Q.: A deep learning approach to network intrusion detection. IEEE Trans. Emerg. Top. Comput. Intell. **2**(1), 41–50 (2018)
20. Yang, Y., Wu, L., Yin, G., Li, L., Zhao, H.: A survey on security and privacy issues in internet-of-things. IEEE Internet Things J. **4**(5), 1250–1258 (2017)

Breast Cancer Detection Using Deep Convolutionnal Neural Networks and Raspberry Pi 3 Board

Iman Elawady, Abdel Kamal Bensaada, and El Mostafa Litim

Abstract Approximately 1 in 8 women (13%) will be diagnosed with invasive breast cancer in their lifetime and 1 in 39 women (3%) will die from breast cancer [1]. The high mortality rate can be decreased by early and accurate screening. As the data shows the early detection correlates heavily with low mortality rates. Screening measures and guidelines differ from country to country, but can fit into two big groups, imaging techniques and biopsy. While generally women are guided to fast but less accurate imaging techniques, doctors generally refer to biopsy when the imaging techniques don't give clear results. This paper uses convolutional neural networks, trained on a biopsy dataset neatly run on single board computer (Raspberry Pi) to advance computer aided diagnosis which would help doctors get a second opinion without referring to multiple doctors, doctors conjointly provide high false positive rates wherever healthy women area unit subject to false diagnosing, and might suffer from psychological and economic issues. The doctor would introduce some pictures of the biopsy cells to the system, which would return a report for the doctor, giving them estimation that the patient has malignant cells (deadly cancerous cells) or benign cells (not cancerous). Pathologists nowadays have 85% accuracy; creating a system that would increase the accuracy to 90% and decrease the false negatives would greatly increase early detection and survival chances for the patients.

Keywords Breast cance · Image processing · Deep convolutionnal neural networks · Raspberry Pi 3 board

I. Elawady (✉) · A. K. Bensaada · E. M. Litim
Department of Electrical Engineering, École Nationale Polytechnique d'Oran-Maurice Audin, Oran, Algeria
e-mail: iman.elawady@enp-oran.dz

E. M. Litim
e-mail: elmostafa.litim@enp-oran.dz

I. Elawady
Laboratoire de Traitement de L'Information Et Télécommunication (LTIT), University Tahri Mohamed Béchar, Béchar, Algeria

L. Troiano et al. (eds.), *Advances in Deep Learning, Artificial Intelligence and Robotics*, Lecture Notes in Networks and Systems 249,
https://doi.org/10.1007/978-3-030-85365-5_10

1 Introduction

With the improvement of computer hardware: more precise and faster processing units, humans have access to faster computers. Which led to new software advancements thus creating more innovative uses for computers. One of these is the development of Artificial Intelligence (AI) which is becoming more prevalent each day in every area of our lives. When used combined with large medical datasets, artificial intelligence can help increase accuracy, and precision and reduce false positives as well as time needed by doctors as it will rise only the suspected cells. In the last five years, the artificial intelligence (AI) revolution in computing, driven mostly by the improvement of deep learning and convolution neural networks, has also pervaded the field of automated breast cancer detection. Although artificial intelligence has come a long way it is still not clear how their performance compares to that of breast screening radiologists in the real screening realm, which can only be evaluated during large-scale screening trials [2]. Computer aided diagnosis (CAD) these days is used concurrently by the radiologist during his/her interpretation of the mammographic exam. Specifically, once the radiologist identifies a finding of concern and is deciding if it is suspicious enough to be recalled or not, the opinion of the AI may be requested when the doctor cannot arrive at a decision. And this is the consequence of the limited availability of homogeneous datasets, which vary from the equipment used: cameras; lens; focus; types of encoding (Gray/RGB, bits per pixel, data format); number of patients and other errors introduced by human intervention [2, 3]. These inconsistencies lead to greater false positives and false negatives. Even though we can not remove all the inconsistencies from our datasets some medical laboratories have tried to create large datasets to help researchers develop accurate and precise algorithms and models to detect and diagnose breast cancer. In this paper we are concerned about building a system that can helps the doctors to detect breast cancer, to achieve this aim our paper is organized as follow in the first part we provide some definitions about breast cancer and Convolutional Neural Networks then we will move to the Implementation and Model creation finally evaluation and results with some perspectives.

1.1 Breast Cancer

Breast cancer is a group of diseases in which cells in breast tissue that change and divide uncontrolled, typically resulting in a lump or mass. Most breast cancers begin in the lobules (milk glands) or in the ducts that connect the lobules to the nipple (Figs. 1 and 2).

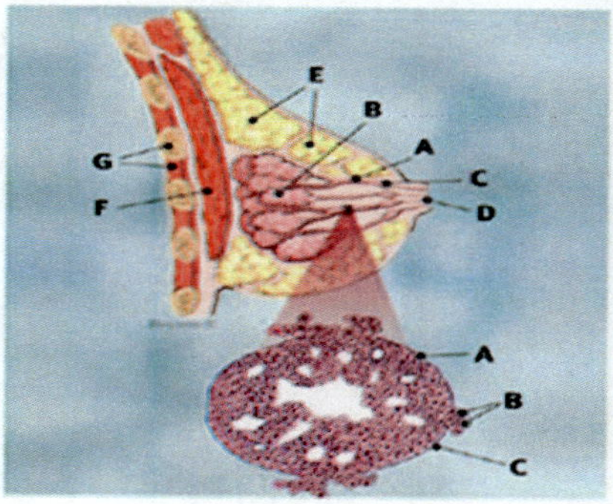

Fig. 1 Normal breast with Ductal carcinoma in an enlarged cross-section

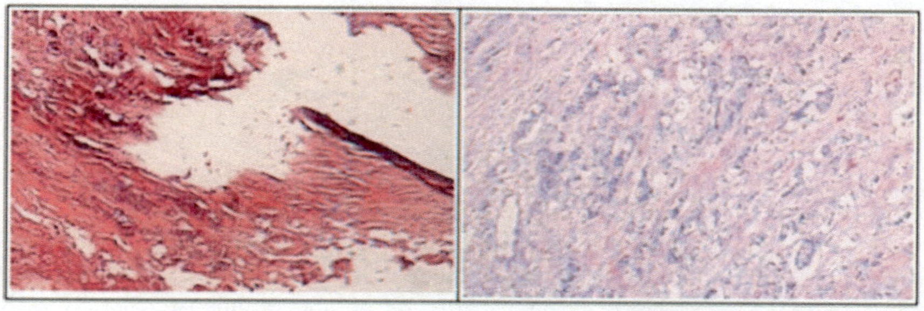

Fig. 2 Examples Ductal carcinoma (Left) Lobular carcinoma (Right) [4]

1.2 Convolutional Neural Networks

A Convolutional Neural Network (ConvNet) is a deep learning algorithm which can take in an input image, assign trainable weights and biases to various aspects/objects in the image and be able to differentiate one from the other. The per-processing required in a ConvNet is much lower as compared to other classification algorithms. While in primitive methods filters are hand-engineered, with enough training, ConvNets have the ability to learn these filters/characteristics [5]. The input image passes through multiple convolution (spatial filters) layers, with multiple types of filters (Up, Down, Diagonal etc.…), the resulting outputs from these convolution layers maybe are funneled through a pooling layer, which is responsible for reducing the spatial size of the convoluted Feature. This leads to decrease the computational power required to process the data due to the dimensional reduction. There are two types of Pooling:

- Max Pooling: it returns the maximum value from the portion of the image covered by the Kernel.

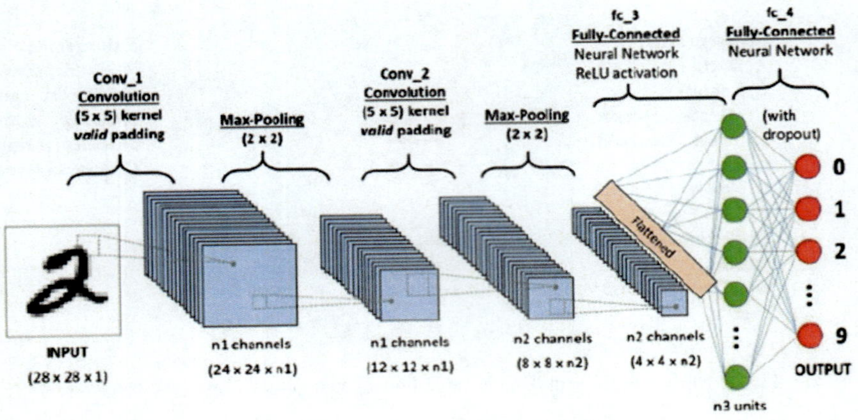

Fig. 3 An example of a ConvNet tasked with recognizing numbers [5]

- Average Pooling: it returns the average of all the values from the portion of the image covered by the Kernel.

The combination of convolution layers and pooling layers may be used several times in the same model depending on the data and the size of the input.

After the size of the data is considerably manageable, the output of the last pooling is connected to a Fully Connected network, these last layers gradually decrease the number of neurons in each layer from the number of inputs (Generally we flatten the data to ease the access) to the number of the desired classes (Fig. 3).

VGGNet
In our paper we are going to use the VGGNet to reduce the number of parameters in convolutional layers. VGGNet only uses 3 × 3 kernels with different number of kernels to replicate the same effect for larger kernels with a great reduction in trainable parameters, which led the creators to the addition of layers that do convolutional layers and get better accuracy and precision, also the model is more robust to overfitting.

The resulting model when fully trained is only around 25 MB, which is very effective on embedded systems where memory and processing power are limited.

1.3 Work Environment

Before moving to the implementation and result section, we present the work environment, both hardware and software. Generally we used:

Raspberry Pi 3 Model B+: because of low cost, availability and the main reason; its specifications are close enough to the implementation requirements.

GPIO (General Purpose Input/Output): are input/output pins that allow the communication with other circuits which we are going to use in our project.
LCD (Liquid Crystal Display): screen so the user knows the state of the program, which leads to ease of use and removes some pre-mature results and bugs.
RGB LED: Like the LCD the RGB (Red Greed Blue) LED (Light emitting Diode) is used as a visual indicator for identification program. The colors identify the following states of the product:

- RED color means that the system is plugged into power and it's waiting for data to start processing it.
- Orange color means that the system is running the code on the data in order to create the results file.
- Blinking Orange means the system didn't complete running the code due to some errors. The errors can be read through the errors file.
- Green color means the code run successfully and the results file was created.

Push Button: Used to start the code when the data is inserted.
Google Colaboratory: Google Colab is used in this project to do the "heavy lifting", in other words it's used to train our model on the Google Cloud without having to bother using the personal computer or Raspberry Pi limited specifications.
Python: The main reason we don't have to use the Raspberry Pi for training the model; limited memory and processing power when compared to system specifications for Google Colab session.
Libraries used: OpenCV, NumPy, TensorFlow, Keras, Pickle module, MatPlotLib.
Breast cancer datasets is taken from:

- Digital Mammography Dataset [6]
- Breast Cancer Histopathological Database (BreakHis) [4]

1.4 Implementation and Results

The schematics and PCB design were created by AutoCad Eagle (v4.2.2).To communicate data to the LCD screen we need to connect the relevant pins of the LCD to the Raspberry Pi GPIO pins that could send data. Only 4 pins of data bits were used (DB4, DB5, DB6, DB7), the other data bits were connected to the ground (GND), as the R/W bit as we only need to write data we needed to connect it to the ground. The VO pin which is responsible on the contrast level was connected to variable resistance to regulate the contrast level. The power pins were connected appropriately. The RGB led (Green, Red, Blue) pins were connected to the GPIO pins 11, 13, 15which correspond to GPIO 17, GPIO 27, GPIO 22. Safety resistors were used to protect the leds from high currents. The K pin was connected to the ground. A push button was connected to RPI and the input pin was connected on the pin 7 which correspond to GPIO 4 (Fig. 4).

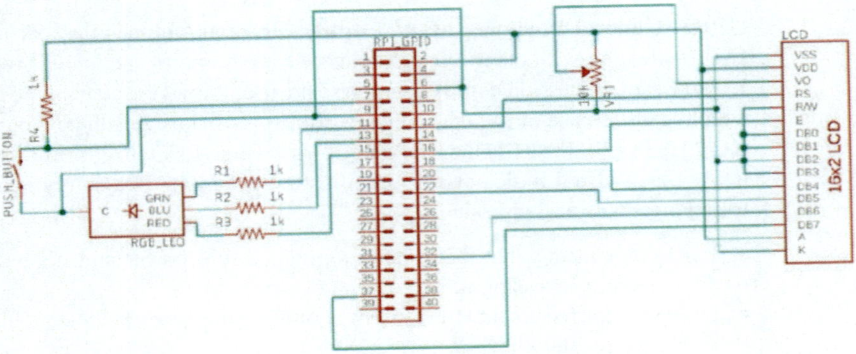

Fig. 4 The project schematics

Printed circuit board: Two layers were used to create the print circuit board. It has a 7.5 × 7.5 cm^2 surface (Fig. 5).

Model

The model used in this implementation is a variation of VGGNet mode. It's a smaller model due the small dataset and mainly while the standard VGGNet16 and VGGNet19 where created to classify 1000 classes our system only needs to classify 2 classes, Benign and Malignant. The model consists of only 6 convolution layers, all the convolution layers have a 3 × 3 size, stride of 1 and same padding, the only difference is the number of filters in each layer (the number of the maps created 32, 64, 64, 128, 128, 128 respectively). It also has only 3 pooling layers with a 2 × 2 kernel which helps decrease the number of trainable parameters to less than 10 millions (Fig. 6).

The model uses the replication of larger convolution kernels with smaller kernels, as it uses two 3 × 3 kernels to create a 5 × 5 kernel and three 3 × 3 kernels to create

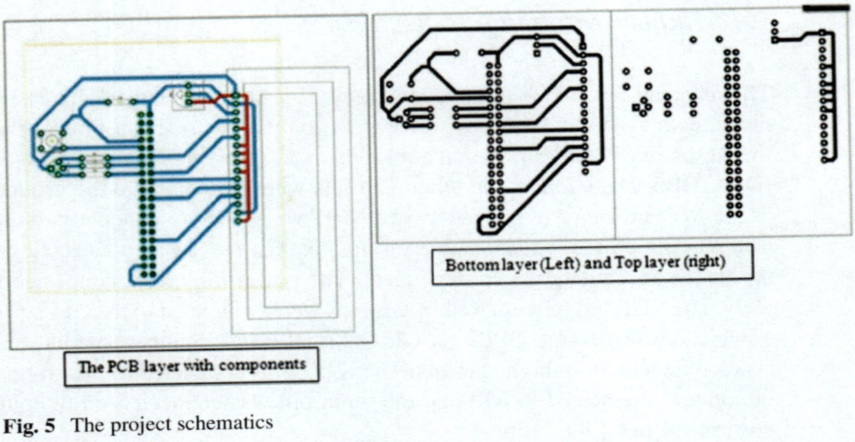

Fig. 5 The project schematics

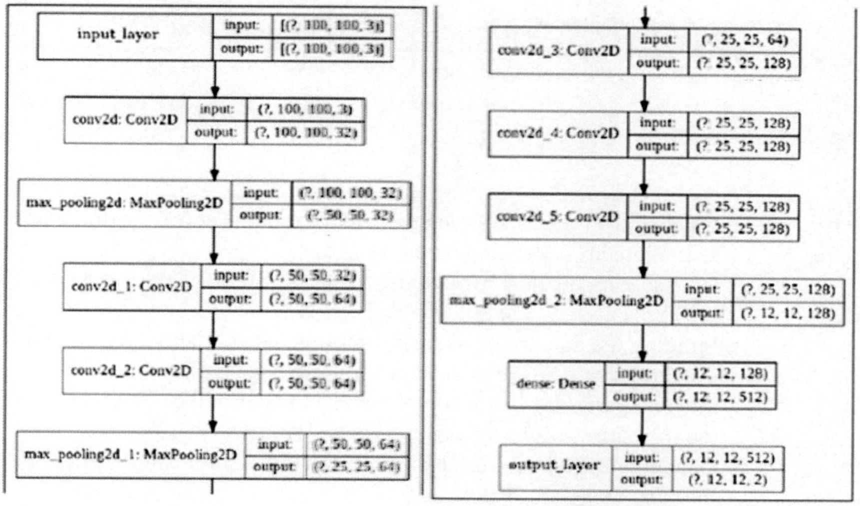

Fig. 6 A representation for the model created

a 7 × 7 kernel. This development decreases the number of needed weights by: 28% for a 5 × 5 kernel and 45% for a 7 × 7 kernel. This helps speed up the training and testing processing by a lot when considering the high number of iterations taken in each training cycle.

In this model two optimization methods were used:

- Batch Normalization
- Dropout

The use of these two techniques made training faster and the model in general more robust and should in theory have more patterns than a non-optimized mode.

Training

After the model was created, the long and heavy task of training the model is next step, Google Colab was used to speed the training. The data used in this implementation was the Break His dataset. The dataset was partially used as only the 40× magnification factor was used as it shows a better view for the cells as the other magnification factors show only partial cells and thus would make the system take longer to converge, the other magnification factors can be virtually created using image augmentation. The available images were divided into two groups Training and Evaluation, where we use them for different steps, the table bellow shows the distribution (Table 1).:

We will talk about the validation test later, for now we will explain the training. It was done by using the model stated in the previous section and was done following these steps:

Table 1 The distribution of training and evaluation data

Data	Benign	Malignant	Total
Training	496	1252	1748
Evaluation	129	118	247
Total	625	1370	1995

1. Loading the images, resizing them to a fixed size and assigning a label (benign or malignant) depending on the image.
2. Normalizing the data by dividing it over 255 making the input a float between [0, 1].
3. Splitting the data into two parts: training and validation, the validation data is used to make sure the model converges to a good place, after splitting the data into two parts, the vector containing the images will be randomized to increase the robustness of the system.
4. Changing the labels from benign and malignant to 0 and 1 (this step is called Serialization)
5. Augmenting the images, where random effects (Rotation, Shifting, Zooming, Flipping etc.…) are applied to some images in the data, these random effects are useful to make our system more robust where the data is relatively small, and to reduce the chance of both non-convergence and over-fitting. As images in small datasets and especially in computer aided diagnosis look similar and the added random effects can help recreate the different real life effects like zooming, cropping and flipping.
6. Building our model using the parameters needed: the size of the image and the number of classes that need to be differentiated from each other.
7. Setting the training method, here the SGD (Stochastic Gradien Descent) with momentum was used. This method tries to estimate the change (gradient) to each weight from a randomly selected subset of data (batch) and the previous batch hence momentum which gives it the "memory" bias, this bias would help in increasing accuracy, in Keras it takes an initial learning rate (INIT_LR) and momentum (decay) can be set to small constant decay = INITLR/EPOCHS where Epochs is the number of loops the training repeats.
8. Setting the metrics to which the model should consider important, for example Accuracy (ACC), False Positive Rate (FPR), False Negative Rate (FNR) etc.…
9. Compiling the model using the already built model waiting to be trained, training method, batch size, INIT_LR and the number of stages (EPOCHS)
10. The model would introduce a batch (a part of the training data), evaluate results depending on the metrics, calculate the gradient (using SGD), update the weights of the neurons and repeat it until it's done with all the training data. After that using the validation data it will update the weights again. This means it finished an Epochs.
11. The compiled model will repeat Step 10 until it finishes the number of EPOCHS set by the user.

12. After it finishes with the training, the program saves the model, the corresponding labels and graphs of the evolution of the metrics the user deemed important.

From the training steps we can detect a fixed number of training parameters, changing these parameters will change the resulting model in terms of the metrics, these parameters are:

- The size of input images: Height and Width and Depth, example: 100 × 100 × 3
- Test (validation) size, step 3, example: 30%
- Initial learning rate, step 7, example: 0.001
- Number of training stages, step 9,
- Batch size, step 10

After fixing these training parameters the model will be trained and will result into a fully trained model.

Evaluation

In this section we will talk about how the evaluation of a fully trained model was done. After saving the model, another code was used to upload the evaluation data talked about in the previous section, this data was never introduced to the model before and thus the model can be judged objectively. After introducing the data to the model it is then judged by some basic metrics which help us calculate other more advanced ones. The basic metrics can be summed up into 4:

- True positive (TP): a true positive sample is when the system predicts that the introduced image is malignant and it is in fact malignant.
- True Negative (TN): a true negative sample is when the system predicts that the introduced image is benign and it is in fact benign.
- False Positive (FP): a false positive (or false alarm) sample is when the system predicts that the introduced image is malignant but it is in fact benign.
- False Negative (FN): a false negative sample is when the system predicts that the introduced image is benign but it is in fact malignant.

A perfect model would have 0 false negatives and 0 false positives.

The number of these basic stats is counted and can be used to calculate more advanced metrics which would help in judging the model. The advanced stats are:

- True positive rate (TPR): also called sensitivity [7] and recall, it measures the proportion of positives that have been correctly identified, it's calculated with this formula $TPR = TP/P = TP/(TP + FN)$
- True negative rate (TNR): also called specificity[7], it measures the proportion of negatives that are correctly identified, it's calculated with this formula: $TNR = TN/N = TN/(TN + FP)$
- Precision: also called positive predictive value (PPV), it answers the following question: if the test was positive, how likely is it that the patient really has the disease? it's calculated with this formula: $PPV = TP/(TP + FP)$

- Negative predictive value (NPV): this metrics answers the following question: if the test was negative, how likely is it that the patient really is healthy? it's calculated with this formula: NPV = TN/(TN + FN)
- Accuracy (ACC): can be defined as how often the model produces true results, it is easily calculated when knowing the already discussed basic metrics with this formula: ACC = (TP + TN)/(P + N) = (TP + TN)/(TP + TN + FP + FN)
- F1-score: The F1 score is the harmonic mean of the precision (PPV) and sensitivity (TPR) as this metric tries to balance between them it is similar to accuracy but it takes in consideration an uneven class distribution (large number of actual negatives)[8], it's calculated with this formula: F1 = (2 × TPR × PPV)/(TPR + PPV)

These metrics are not the full list, as others compliment these such as:

- False positive rate (FPR): FPR = 1 − TNR
- False negative rate (FNR): FNR = 1 − TPR
- False discovery rate (FDR): FDR = 1 − PPV
- False omission rate (FOR): FOR = 1 − NPV

After the code goes through all the images in the evaluation data, it will calculate the metrics stated above and then we will judge if the model is good or not.

The system works as follows

1. The doctor would take several images of the cells taken from a biopsy procedure (it works better with Surgery Open Biopsy)
2. He would introduce these images into the Raspberry Pi via a USB or through SSH communication, the images will be placed in a special folder.
3. The Pi would run a prediction algorithm on all the images present, and return a value through the LCD and would return error information if problems are raised through the LCD and the RGB led.
   a. RED color means that the system is plugged into power and it's waiting for data to start processing it.
   b. Orange color means that the system is running the code on the data in order to create the results file.
   c. Blinking Orange means the system didn't complete running the code due to some errors. The errors can be read through the errors file.
   d. Green color means the code run successfully and the results file was created.

Test

The input parameters are selected randomly just to test the system, our system shows a good results, also the possibility to get a high accuracy compared with existing systems (Pathologists nowadays have 85% accuracy) According to the presented results it's seems that we could achieve more than 85%. The figures below shows some of tested samples with different accuracy level depending on the input parameters (Fig. 7).

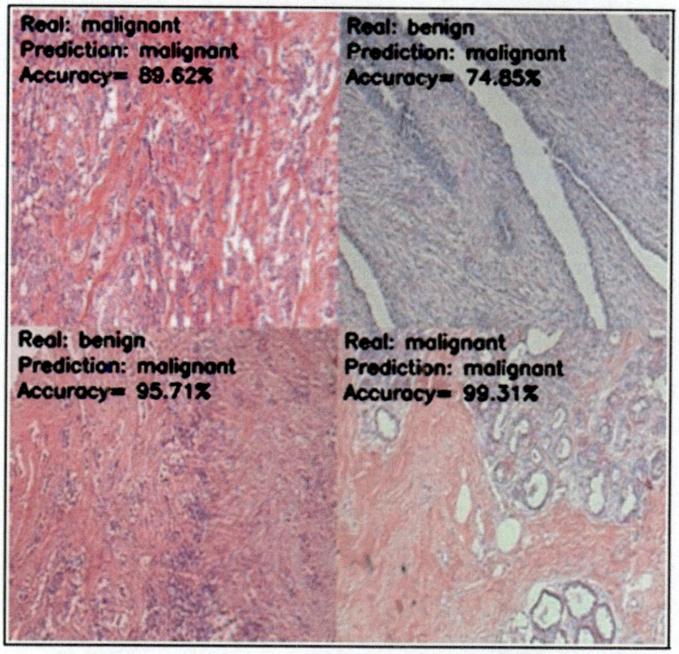

Fig. 7 4 tests samples and the results with the accurate of the prediction

2 Conclusion

A diagnostic system in the real time has been implemented and tested successfully, the proposed system shows a high performance with limited hardware systems, our next step is to study the input parameters wisely to get the highest accuracy could be found additionally, making some investigation to obtain better parameters of ConvNet algorithm. Also if we implement our system with high performance hardware for image processing like FPGA, could enhance our diagnostic model with big set.

References

1. American Cancer Society Journal, CA: A Cancer Journal for Clinicians, Breast Cancer Facts & Figures, , 2019–2020
2. Sechopoulos, I., Jonas T., Ritse M.: Artificial intelligence for breast cancer detection in mammography and digital breast tomosynthesis: state of the art. Semin. Cancer Biol (2020)
3. Rodríguez-Ruiz, A., et al.: Detection of breast cancer with mammography: effect of an artificial intelligence support system. Radiology **290**, 305–314

4. Spanhol, F., Oliveira, L.S., Petitjean, C., Heutte, L.: A Dataset for breast cancer histopathological image classification. IEEE Trans. Biomed. Eng. (TBME) **63**(7), 1455–1462 (2016)
5. Saha, S.: A comprehensive guide to convolutional neural networks—the ELI5 way. Towards Data Science. Towards Data Science (2018)
6. https://www.bcsc-research.org,https://www.bcsc-research.org/data/mammography_dataset. Last Accessed 8 Sept 2020
7. https://en.wikipedia.org/wiki/Main_Page,., https://en.wikipedia.org/wiki/Sensitivity_and_specificity. Last Accessed Jan 2019
8. Koo PS.: Accuracy, Precision, Recall or F1? Towards Data Science. Towards Data Science, 15 (2018)

An Improved Approach for Removal of Salt and Pepper Noise in MR Images

Monika Jyotiyana, Nishtha Kesswani, and Anupam Agarwal

Abstract Removal of the noise entirely is the most intricate process, as it sometimes ruins the visual excellence of image along with its details. So, the exclusion of noise is imperative for the better quality of images. Noisy pixels also reserves the memory; that space is unnecessary reserved for senseless data. Noise added in an image during image acquisition mostly. In this manuscript, we present an efficient method for removing salt and pepper noise using the modified weighted mean filter. It includes noise identification and removal processes from the Magnetic Rescnance Images. And reproduce an enhanced de-noised image. In the proposed method, we use the cosine rule for detecting the noisy pixel and replaced by their mean value within a window size 3×3. Similarly, we applied the proposed methods on various filters like alpha trimmed filter, mean, mode and median filter and all the filters give promising results. Experiment results indicate efficiency over other approaches in removing salt-and-pepper noise.

Keywords Salt and pepper noise · Denoising · Additive noise · Weighted mean filter

1 Introduction

The digital image is habitually corrupted by impulse noise during image acquisition and diffusion. Impulse noise is namely known as salt-and-pepper noise, and it spotted with the extreme values, which are minimum and maximum value in the dynamic range [1] and diagnose in the form of black and white specks that's why known as salt-

M. Jyotiyana (✉) · N. Kesswani
Central University of Rajasthan, Ajmer, Rajasthan, India

A. Agarwal
Anand ICE, Jaipur, Rajasthan, India

L. Troiano et al. (eds.), *Advances in Deep Learning, Artificial Intelligence and Robotics*, Lecture Notes in Networks and Systems 249,
https://doi.org/10.1007/978-3-030-85365-5_11

and-pepper noise. It ruins an image's quality and appearance, therefore effectively removing impulsive noise from MR image, stand for accurate diagnosis and research purposes.

Noise is a redundant part of an image. It lessens the quality of an image by disturbing the value of the pixels. Most images are distorted by the random addition of noise in transmission channels. Exclusion of noise has always remain as the main task in image processing unit [1–7]. There are several grounds for the preface of such noises in an image. Noises can be commenced in images by both ways, which could be made possible through natural or artificial ways. Likewise, for example, they are electronic sensors, temperature, fog, movement of the object, and other reasons like lapse in data communication, the lapse in analog to digital conversion, malfunctioning of pixels in electronic device sensors, inadequate memory locations, dead pixels time-based lapse during digitisation etc., remains to be the significant sources of bringing the noise in an image. Hence, the main function of attaining the finest quality of an image is always a necessity of noise exclusion.

Salt and pepper noise is the most ordinary noise in an image. It is a trademark of impulse noise and also epitomises as an intensity spike. In the case of salt and pepper noise, the pixel's value becomes maximum or minimum, i.e. either 0 or 255. The value of salt noise is 255 (signified by white colour) and pepper noise is 0 (signified by black colour) in case of uint8 image. There are many existent filters to minimise the effect of additive noises, such as mean, median, mode and alpha trimmed filters and weighted median filter (WMF), and many more as the mentioned filters has some of the main drawbacks in affecting the noise and virgin pixels which are leading towards the increase in time and space complexity. The changing of the value of virgin pixels also affects the information carried by the image signal.

This research has taken up a significant task in trying to come out with an effective solution by resolving the issues mentioned earlier and introducing a new form of approach that would not affect the value of the original pixel and help retain the authenticity of the image. The method taken up for the study is the curvilinear triangle method so, we give a name to our proposed method as C-trigon method. In the RGB colour images, RGB beams combine to form a new colour, and all these three beams merge and form a structure like a curvilinear triangle [8]. And this is the groundwork for the proposed C-trigon method. The function of the approach would emphasise on detecting the noisy pixels and getting repaired, without involving the original pixels. It maintains the originality of the image as much as possible.

This manuscript is systemised as follows. Section 2 describes the background. Section 3 shows dataset and analysis description. Section 3.1 includes methodology. Section 4 experiments and results and Sect. 5 concludes the manuscript.

2 Background

In last decades many methods [9–16] have been proposed to eliminate the impulse noise. Non-linear filters perform more proficient than linear filters, and the extensively used median filter is easy to understand and efficient too. However, the performance of the median filter is inversely proportional to noise density. Thus many modified median filters have been proposed [12, 13]. Weighted median filters [13] used the neighbor pixel, so it improves the quality of median filter even though at high density noise still affects the performance of the filter [9–12, 17].

The adaptive median filter can enhance median filter performance [15], because of the adjustable filtering window size to the noise content, and such filters are applied homogeneously throughout the image. Hence required information may get loss. The switching median filters can overcome this problem, it consist of two stages i.e. noise detection and noise removal, respectively. Example of switching median filter is an efficient adaptive switching median filter [18]. Later on for removal of high-density noise an improved switching median filter introduced, which constructed on weighted median filter [19].

All these switching based median filters perform better than the standard median filter, at high-density noise image quality degrades, because they rely on local statistics within small neighbour pixel to remove impulse noise. Hence leads to destroying the details of an image. Buades et al. [20] introduced Non-local Means Filter (NLM) filters, the decision based non-local means filter performs more efficient than non-local means filters, DNLM having more computational complexities [21]. As a solution, switching non-local means filter (SNLM) has been introduced to remove impulse noise, due to using big size filtering window leads to ruin the filtered image blur. An improved median filter (IMF) [19] uses the minimum absolute value of four convolutions acquired by 1D Laplacian operators. These above-stated methods enhance the performance of the median filter but require some parameters and more processing time. Furthermore, for removing the impulse noise, fuzzy filtering techniques and total variation methods [22] have been proposed.

In 2014, Danni et al. discussed that eleven new denoising filters proposed for 15 years and, NLM and BM3D are the most significant ones. In this paper, they discussed the HOSVD method which reduces the amount of MR data which were not useful 3D images, and are already having the magnificent information in it so that during the process of denoising, they removed the data size which is quite useful during data transmission.

In the year 2014 Mohan et al. discussed the various denoising methods of 3D images, such as the spatial and temporal filtering, anisotropic diffusion filter, non-local mean filtering, combination of domain and range filtering techniques, wavelet transform, curvelet transform, contour transform, and many statistical approaches [23].

In the year 2017, Zhang et al. have discussed the new technique removal of noise using the tool of convolution feed-forward neural network and compared the DnCNN technique with existing techniques ground-truth and CBM3D. DnCNN can perform more effectively in Gaussian denoising, single image super-resolution and JPEG image deblocking [24].

Ali discussed various denosing filters in his book [25] he discussed adaptive wiener filter and proposed the modified method for modifying the median filter, which gives better results than a median filter.

Panigrahi et al. proposed a novel filter using NLM and guided image filter. The proposed method, combined with hard thresholding in the curvelet domain to preserve details of an image, edges and texture, is used in guided filters. It can use for colour as well as grayscale images [26].

Many methods are recently proposed for denoising, like CNN [27], DNN and other deep learning models. Moreover, many researchers working on it as day-by-day new technology invented noise and errors also increases, making it a never-ending research area.

3 Dataset and Analysis Description

We have used MRI data for the experiment, data is freely available on www.adni.loni.usc.edu, this organisation is provide the data for study and research purposes. We have used Alzheimers Disease (AD) patients data for denoising. Noise is unwanted part of any image it is crucial to remove the noise from the image for getting maximum fruitful information. We have used 12 MRI images for the experiment, MR images are larger in size since they carry lots of information in it. For the experiment purpose we choose few images for faster processing. Our system requirements for the experiments are 4 GB RAM intel i5 processor with 7th generation and MATLAB R2016b used.

3.1 Methodology

Non-linear morphological filter can remove the high-density impulse noise, but the performance is not effective. In the morphological filter, the structuring element has a static shape and size, creating or removing essential details of an image. For overcome these problematic issues, we proposed a method. The proposed method uses a noisy pixel detection based on cosine rule. In the noise removal stage, only noisy pixels processed by the modified weighted mean filter.

Phase 1: Noise Detection

The dynamic range of an image is either the maximum value or the minimum value, which lies between (i.e. 0 to N − 1). This can be signified using this equation.

$0 \leq f(i, j) \leq N - 1 \quad \text{and} \quad i \leq M \quad \text{and} \quad j \leq N$

where [M,N] = size (f).

Phase 2: Removal of Noise

Apply the C-trigon method to detect the 0 valued pixels; this will give the position of the affected pixel and application of filter (WMF/alpha median etc.) will repair noisy pixel.

Phase 3: Apply Filters

We have applied the mean, mode, median, alpha-trimmed and weighted median filter using the C-trigon approach. And found a remarkable improvement in each of the filters.

Phase 4: Calculate PSNR and SSIM

In last stage we calculate PSNR and SSIM and compare the different methods.

4 Experiments and Results

In the existing methods, most of the filters are implemented homogeneously all over the image. Thus removal of necessary details of an image is usual. So to the solution of the problem, in the proposed C-trigon method, we will consider the small size filtering window and only pick the noisy pixels from an image. We used mean, mode, median, WMF and alpha trimmed filter for experiment. The results are shown in Figs. 1, 2 and 3.

In Figs. 1, 2 and 3 we show the results of WMF filter and alpha trimmed filter with proposed C-trigon method as we can easily measure the difference using PSNR and SSIM value that C-trigon method performs much better then exiting WMF and alpha trimmed filter even if we change the parameters like noise density and alpha value in that case C-trigon method performance is remarkable.

In Figs. 4, 5 and 6 we show the results of Mean, Mode and Median filter and results of proposed C-trigon method with these existing filters. As we can easily find out using PSNR and SSIM value that C-trigon method performed much better then exiting filter even if we change the noise density.

As we can see in Fig. 7 that proposed C-trigon method performed great as compare to existing filter with different noise density it proves that proposed method have capability of removing the noise better then existing methods.

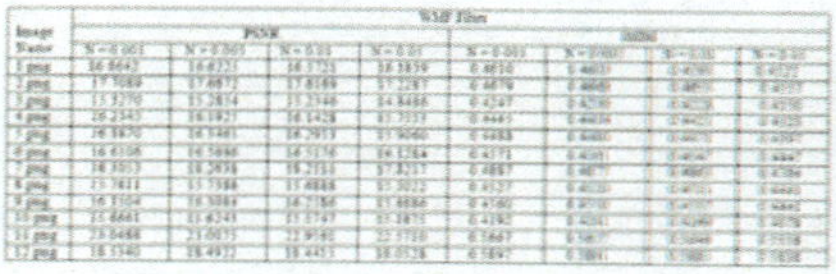

(a) WMF filter

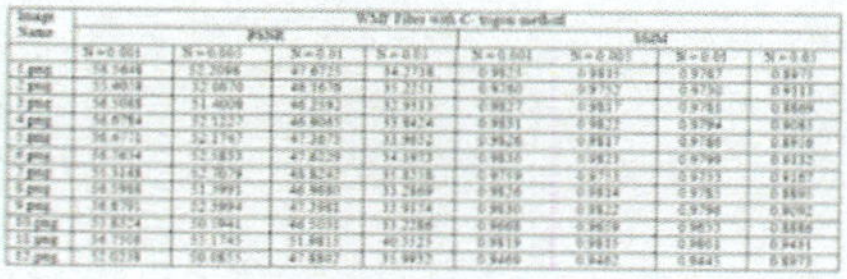

(b) WMF with C-trigon method

Fig. 1 **a** Shows the results of WMF filter and **b** shows the result of WMF filter with proposed method

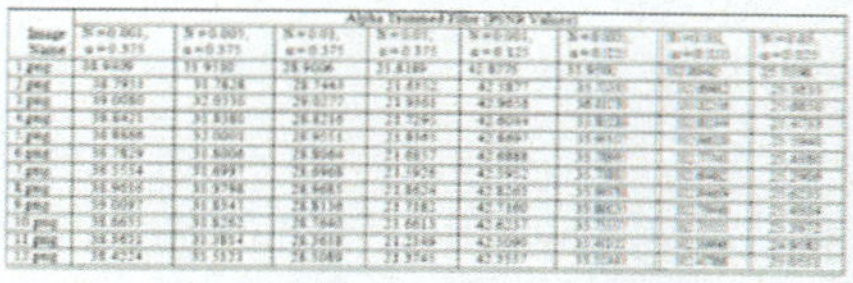

(a) Alpha trimmed filter PSNR values

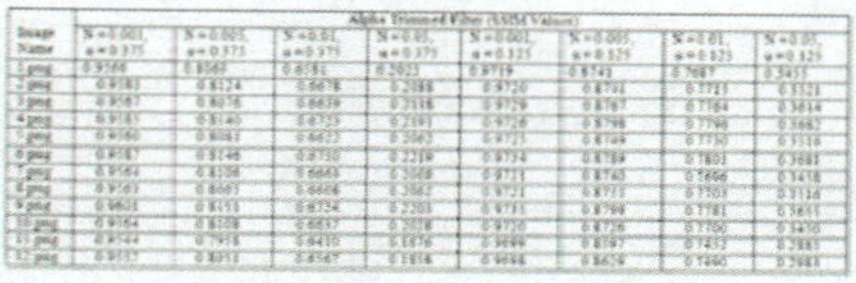

(b) Alpha trimmed filter SSIM values

Fig. 2 **a** Shows the results of alpha trimmed filter PSNR value and **b** shows the result of alpha trimmed filter SSIM value

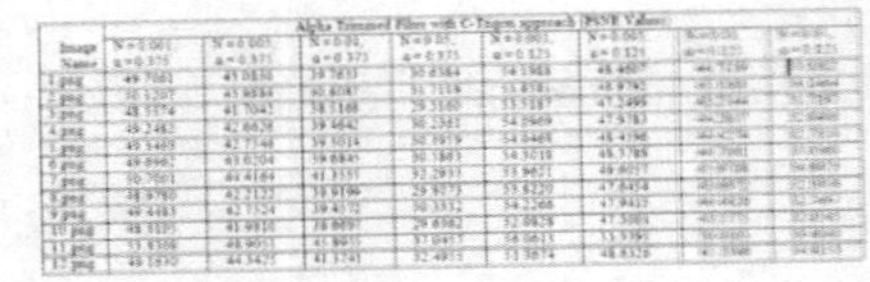

(a) Alpha trimmed filter PSNR values with C-trigon method

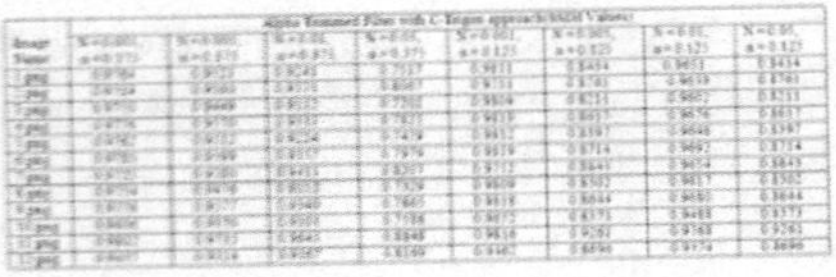

(b) Alpha trimmed filter SSIM values with C-trigon method

Fig. 3 **a** Shows the results of alpha trimmed filter PSNR value with C-trigon method and **b** shows the result of alpha trimmed filter SSIM value with C-trigon method

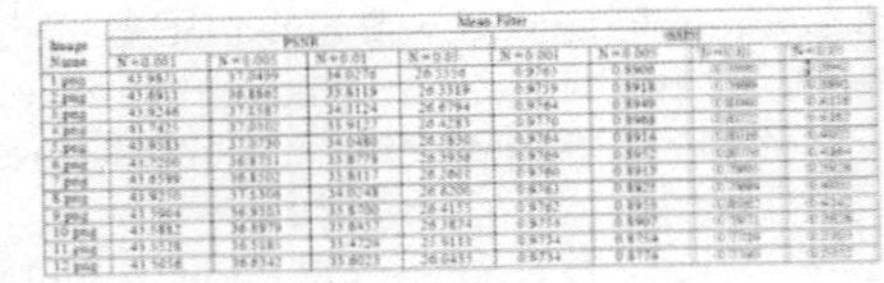

(a) Mean filter

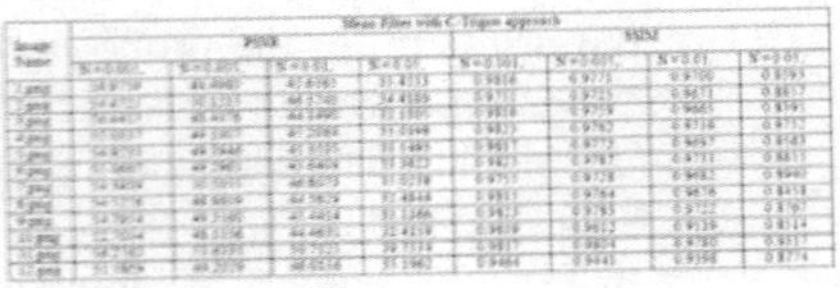

(b) Mean with C-trigon method

Fig. 4 **a** Shows the results of mean filter and **b** shows the result of mean filter with proposed C-trigon method

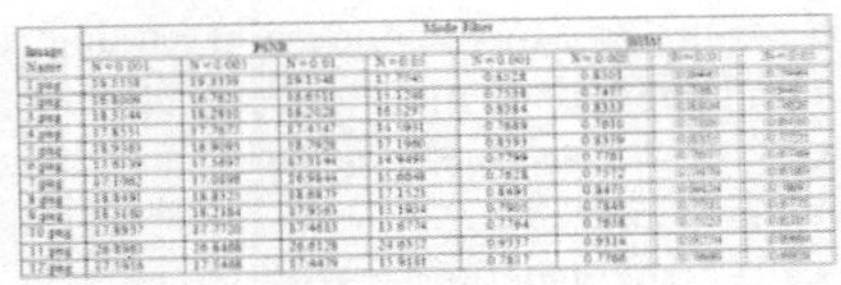

(a) Mode filter

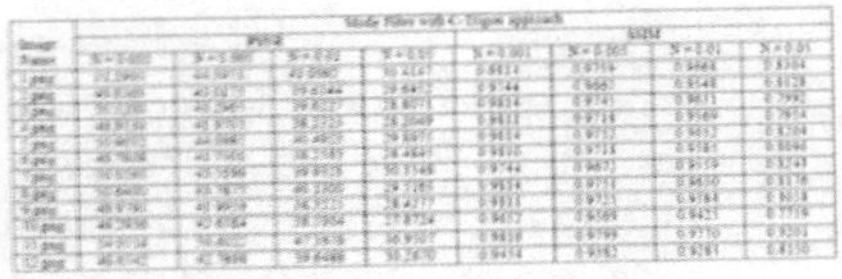

(b) Mode with C-trigon method

Fig. 5 **a** Shows the results of mode filter and **b** shows the result of mode filter with proposed C-trigon method

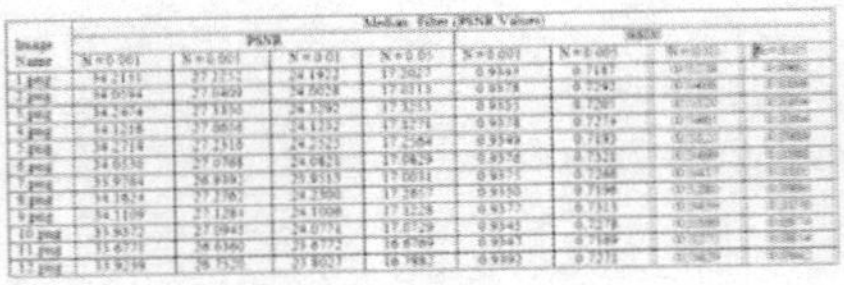

(a) Median filter

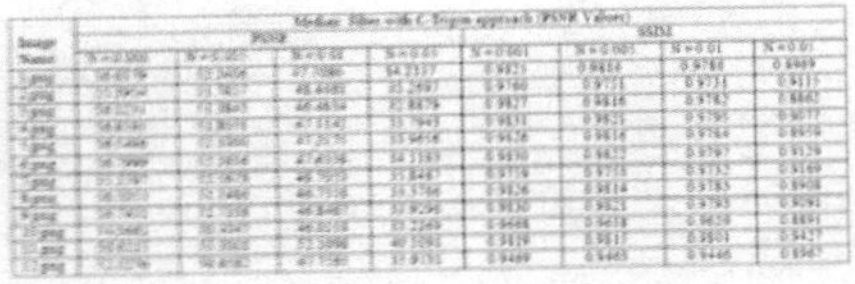

(b) Median with C-trigon method

Fig. 6 **a** Shows the results of median filter and **b** shows the result of median filter with proposed C-trigon method

5 Conclusions

Noise is unwanted part of any digital image it usually exist in every image. In this manuscript we proposed a novel method for removal of high density impulse noise from digital image. A very simple noise detection mechanism is defined in this manuscript, which is not only productive for impulse noise removal while its preserves the originality of an image, by choosing only noisy pixel from the filtering

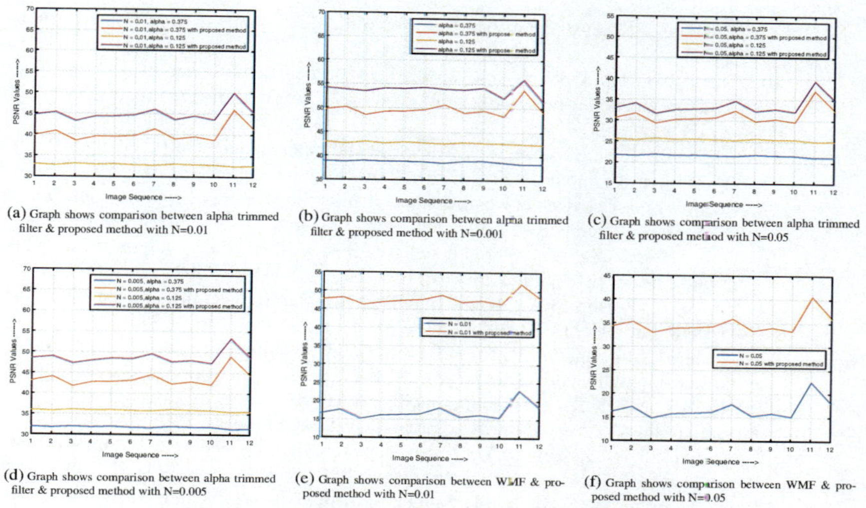

(a) Graph shows comparison between alpha trimmed filter & proposed method with N=0.01

(b) Graph shows comparison between alpha trimmed filter & proposed method with N=0.001

(c) Graph shows comparison between alpha trimmed filter & proposed method with N=0.05

(d) Graph shows comparison between alpha trimmed filter & proposed method with N=0.005

(e) Graph shows comparison between WMF & proposed method with N=0.01

(f) Graph shows comparison between WMF & proposed method with N=0.05

Fig. 7 Graphs **a**, **b**, **c** and **d** shows the comparison between alpha trimmed filter and C-trigon method with noise density (N) = 0.01, 0.001, 0.05 and 0.005 respectively and in graphs **e** and **f** shows the comparison between WMF filter and C-trigon method with noise density (N) = 0.01 and 0.05

window. Results reveal the efficiency of the proposed method. We also perform the experiments using mean, mode and median filter all these filters also gives better results with proposed C-trigon method. The proposed method have a positive advantage that it does not require any specific parameter to be tuned.

References

1. Gonzalez, R.C., Woods, R.E.: Image processing. Digit. Image Process. **2**, 1 (2007)
2. Gonzalez, R.C., Woods, R.E., Eddins, S.L.: Digital Image Processing Using MATLAB. Pearson Education India (2013)
3. Rani, V.: A brief study of various noise model and filtering techniques. J. Glob. Res. Comput. Sci. **4**(4), 166–171 (2013)
4. Verma, R., Ali, J.: A comparative study of various types of image noise and efficient noise removal techniques. Int. J. Adv. Res. Comput. Sci. Softw. Eng. **3**(10), 617–622 (2013)
5. Aguilar-Gonzalez, P.M., Kober, V., Diaz-Ramirez, V.H.: Adaptive composite filters for pattern recognition in nonoverlapping scenes using noisy training images. Pattern Recogn. Lett. **41**, 83–92 (2014)
6. Singh, P., Arora, A.: Analytical analysis of image filtering techniques. Int. J. Eng. Innov. Technol. (IJEIT) **3**(4), 29–32 (2013)
7. Patidar, P., Gupta, M., Srivastava, S., Nagawat, A.K.: Image de-noising by various filters for different noise. Int. J. Comput. Appl. **9**(4), 45–50 (2010)
8. Brown, W.J., Wilton, D.R.: Singular basis functions and curvilinear triangles in the solution of the electric field integral equation. IEEE Trans. Antennas Propag. **47**(2), 347–353 (1999)

9. Windyga, P.S.: Fast impulsive noise removal. IEEE Trans. Image Process. **10**(1), 173–179 (2001)
10. Chan, R.H., Ho, C.W., Nikolova, M.: Salt-and-pepper noise removal by median-type noise detectors and detail-preserving regularization. IEEE Trans. Image Process. **14**(10), 1479–1485 (2005)
11. Srinivasan, K.S., Ebenezer, D.: A new fast and efficient decision-based algorithm for removal of high-density impulse noises. IEEE Signal Process. Lett. **14**(3), 189–192 (2007)
12. Ko, S.J., Lee, Y.H.: Center weighted median filters and their applications to image enhancement. IEEE Trans. Circuits Syst. **38**(9), 984–993 (1991)
13. Dong, Y., Xu, S.: A new directional weighted median filter for removal of random-valued impulse noise. IEEE Signal Process. Lett. **14**(3), 193–196 (2007)
14. Zhang, S., Karim, M.A.: A new impulse detector for switching median filters. IEEE Signal Process. Lett. **9**(11), 360–363 (2002)
15. Ibrahim, H., Kong, N.S.P., Ng, T.F.: Simple adaptive median filter for the removal of impulse noise from highly corrupted images. IEEE Trans. Consum. Electron. **54**(4), 1920–1927 (2008)
16. Ng, P.E., Ma, K.K.: A switching median filter with boundary discriminative noise detection for extremely corrupted images. IEEE Trans. Image Process. **15**(6), 1506–1516 (2006)
17. Fabijanska, A., Sankowski, D.: Noise adaptive switching median-based filter for impulse noise removal from extremely corrupted images. IET Image Process. **5**(5), 472–480 (2011)
18. Nallaperumal, K., Varghese, J., Saudia, S., Arulmozhi, K., Velu, K., Annam, S.: Salt & pepper impulse noise removal using adaptive switching median filter. In: OCEANS 2006-Asia Pacific, May 2006, pp. 1-8. IEEE (2006)
19. Wang, S.S., Wu, C.H.: A new impulse detection and filtering method for removal of wide range impulse noises. Pattern Recogn. **42**(9), 2194–2202 (2009)
20. Buades, A., Coll, B., Morel, J.M.: A non-local algorithm for image denoising. In: 2005 IEEE Computer Society Conference on Computer Vision and Pattern Recognition (CVPR 05), June 2005, vol. 2, pp. 60–65. IEEE (2005)
21. Zhang, X., Zhan, Y., Ding, M., Hou, W., Yin, Z.: Decision-based non-local means filter for removing impulse noise from digital images. Signal Process. **93**(2), 517–524 (2013)
22. Wang, G., Zhu, H., Wang, Y.: Fuzzy decision filter for color images denoising. Optik **126**(20), 2428–2432 (2015)
23. Mohan, J., Krishnaveni, V., Guo, Y.: A survey on the magnetic resonance image denoising methods. Biomed. Signal Process. Control **9**, 56–69 (2014)
24. Zhang, K., Zuo, W., Chen, Y., Meng, D., Zhang, L.: Beyond a Gaussian denoiser: residual learning of deep CNN for image denoising. IEEE Trans. Image Process. **26**(7), 3142–3155 (2017)
25. Ali, H.M.: MRI medical image denoising by fundamental filters. In: High-Resolution Neuroimaging—Basic Physical Principles and Clinical Applications, pp. 111–124. InTech (2018)
26. Panigrahi, S.K., Gupta, S., Sahu, P.K.: Curvelet-based multiscale denoising using non-local means & guided image filter. IET Image Process. **12**(6), 909–918 (2018)
27. Jifara, W., Jiang, F., Rho, S., Cheng, M., Liu, S.: Medical image denoising using convolutional neural network: a residual learning approach. J. Supercomput. **75**(2), 704–718 (2019)

Dual-Pathway Deep CNN for Offline Writer Identification

Naresh Purohit and Subhash Panwar

Abstract Writer Identification is a challenging problem based on small amount of handwritten text. It is also a foremost research topic in forensic analysis of documents. For this, a convolution neural network (CNN) is proposed to address the writer recognition task. In this work, we propose a dual-path CNN to extract local features maps from given input handwritten document image patches and then classify writer by Softmax loss function. The experiments are done on IAM English language dataset and obtained accuracy of 92.7%.

Keywords Writer identification · Deep learning · Feature extraction

1 Introduction

Handwriting is still an effective way to represent the uniqueness of individual, and plays an essential role in bio-metric identification. In Document Image Analysis and Recognition, the Writer Identification Problem is a stimulating task due to intensive variation of human writing styles. Writer Classification is the process to classify the writer of the queried document. It is categorised into two parts (a) Offline (b) Online. In offline writer identification, we deals with text character, lines and paragraphs in static handwritten scanned images. Writing angle or pressure and speed of writing, these type of information considered by Online writer identification at the same time when it is generated by mouse or light pen devices. The writer identification task is based on text dependent or text independent approaches. The comparison of same data written by different writers in text dependent approach however text independent approaches compare different data content by different writers. Text independent got a broad applicability but do not obtain the same high accuracy as text dependent methods do [1], our proposed model is based on the text independent offline writer identification task.

N. Purohit (✉) · S. Panwar
Department of Computer Science Engineering, Engineering College Bikaner, Bikaner, India

L. Troiano et al. (eds.), *Advances in Deep Learning, Artificial Intelligence and Robotics*, Lecture Notes in Networks and Systems 249,
https://doi.org/10.1007/978-3-030-85365-5_12

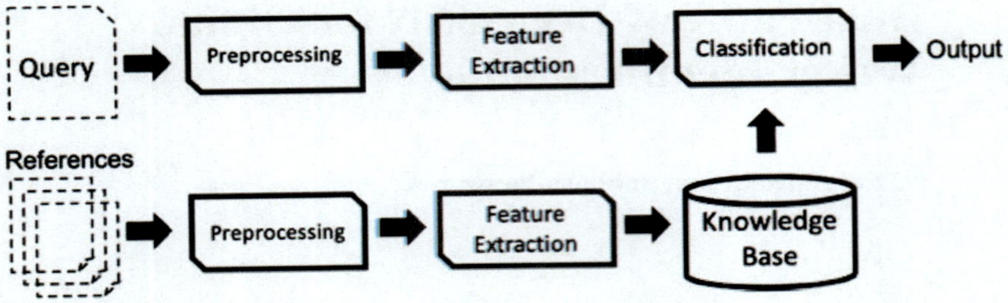

Fig. 1 Basic pipeline of writer recognition system

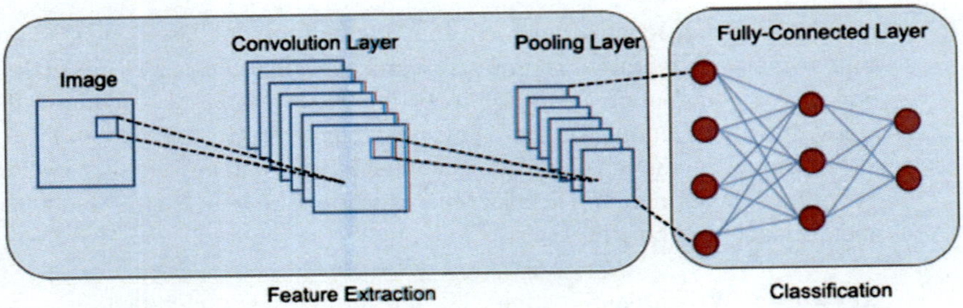

Fig. 2 Basic CNN architecture

Figure 1 shows the basic modules of any writer recognition system which includes pre-processing, feature extraction and classification. Pre-processing consists data cleaning operations like remove noise from image data, segment the image into small patches, normalize image patches and morphological operations. Features of image data is extracted by traditional handcrafted feature based or deep learning based methods. At the last stage, classifiers is used to predict the writer of queried image document. In this paper, we propose Dual-Pathway Deep CNN model, which extract feature maps from original and skeleton image patches. Our proposed model is based on the basic architecture of Convolution Neural Network(CNN). In computer vision, CNN is a deep learning algorithm that can frequently used for recognize and classify features in images. The basic CNN architecture (inspired from LeNet-5) have two main parts. First is for feature extraction and second is for classification purpose as shown in Fig. 2.

Convolution Layer (CL) and Pooling Layer (PL) are the part of feature extraction phase of CNN. CL generates several feature maps by applying different filters on input image. PL down-samples the feature maps generated by CL and maintains the most essential information. After final pooling operation the 2D down-sampled feature maps converted into 1D vectors and passed to fully connected layer for classification process. Finally a probability loss function (Softmax) is used to predict the writer class of test query document.

Accordingly, the rest of the paper is organized as follows: Related work is discussed in section two, The methodology is subsequently explained in section three with proposed data pre-processing and dual-pathway CNN model. Experimental results are then evaluated in section four, and, finally, in section five, conclusion and future work will be summarized.

2 Related Work

In this section, we reviewed several typical offline writer recognition methodologies. The traditional handcrafted feature based methods are used before the era of deep learning based neural networks. In the handcrafted feature based methods, generally we consider texture and shape features of image data. Bulacu et al. [2] used total six different probability distribution functions (PDFs) which is extracted from the handwriting images with texture and allograph level to identify writers on queried documents and got good accuracies with the combination of two texture (Direction Co-Occurrence PDFs and Run-Length PDFs) and one allograph (Grapheme Emission PDF) features. After this in 2011, Jain et al. [3] introduces a new method to improve the performance using K-adjacent segment (KAS) features. They also applied exemplar clustering on the KAS features and used Euclidean distance for writer classification then got 89.6% accuracy rate on IAM dataset. Further the same author in 2014, combined the KAS features with SURF and contour gradient descriptors and improved the accuracy from 89.6 to 94.1% [4]. LoG filter with SIFT descriptors is used by Xiangqian et al. [5] on HIT-MW chienese dataset to extract words regions from documents and they achieved 95.4% accuracy rate.

Deep learning based algorithms usually show better recognition performance compared to hand-crafted feature based methods. The reason is huge amount of data with the large number of features. Convolution Neural Network(CNN) is a deep learning algorithm that can recognize and classify features in images data. The basic architecture of CNN is shown in Fig. 2. In recent years, several CNN deep learning models combined with some operations like VLAD encodings, SVM etc. applied on extracted features and got good accuracy rates for writer recognition. In Xing et al. [6], they used LeNet-5 like CNN structure on IAM dataset with 5 Convolutional Layer(CL) and 3 Max Pooling(MP) Layers to extract feature maps. A another technique proposed by Christlein et al. [7], in which they construct cluster index using clustering of SIFT (Scale-invariant feature transform) key-points extracted from image data, which is further used as image target in their CNN Model. Some of deep learning models can also perform with un-labelled data like Shiming chen et al. [8] proposed a model with weighted uniform label distribution method called Weighted label smoothing regularization (WLSR) for un-labelled data. They used ResNet-50 which is additional flavour of CNN and achieves 99.2% accuracy on CVL dataset.

3 Methodology

In this section, We discuss the proposed pipeline for offline writer identification based on scanned handwritten image patches using CNN deep learning model. The pipeline consists of following stages:

- Pre-process the scanned handwritten image data from IAM Dataset.
- Feature extraction of pre-processed image data using proposed Dual-Pathway Deep CNN Model.
- Writer class prediction using Softmax.

3.1 Preprocessing

Data preprocessing is introduced due to enhance image quality by eliminating unwanted data and improve readability information for constructing suitable image. We used IAM handwriting database in this paper. IAM dataset have handwritten scanned text images from more than 300 writers in form of line and paragraph image. We consider the single text line handwritten scanned image dataset of IAM shown in Fig. 3.

We apply some pre-processing operations on these text line images before passing it to the proposed CNN model. First all text line scanned images convert into single channel, called binarization using Otsu's method. After binarization of text line images, We also find skeleton of that image. Skeleton of image is known as morphological operation as it provides the useful information about shapes in image. Both operations are shown in Fig. 4.

After analysis of all the text line scanned images of IAM dataset, we got that all images with diffrent height and width where for all images width is greater than height. Since the proposed CNN model always take input as squared shape image means that width and height of input image should be same. Therefore, to deal with non-squared text line image, there is a patch cropping strategy we used. In the patch cropping technique there are following two phases as shown in Fig. 5.

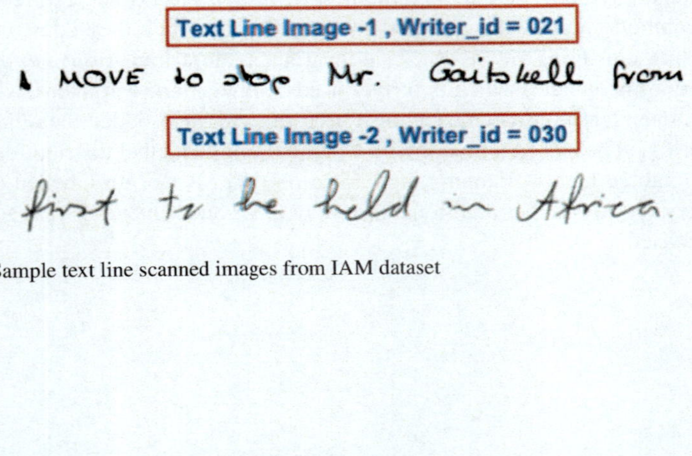

Fig. 3 Sample text line scanned images from IAM dataset

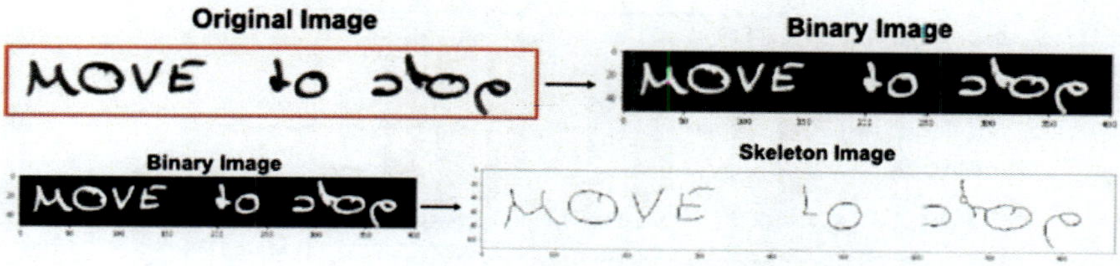

Fig. 4 Binarization and Skeleton operations on text line scanned image

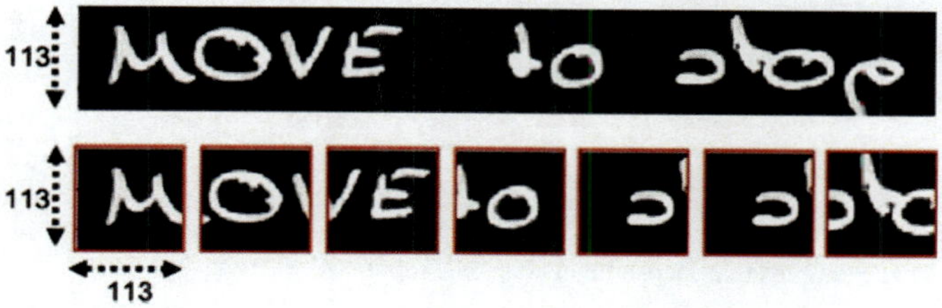

Fig. 5 Sample text line scanned images from IAM dataset

- Resize all text line images to fixed height 113px and their corresponding width with maintaining aspect ratio.
- Crop the image patches of 113 × 113 from that resized image.

For a single text line image, the patch cropping process is applied on their binarized and (binarized+skeletonized) images. Therefore, in the pre-processing stage, every single text line image have many 113 × 113 single channel skeleton image patches. These squared image patches passed to the proposed CNN model as input.

3.2 Dual-Pathway Deep CNN

The primary network structure of our CNN model is inspired from AlexNet structure [9]. The proposed CNN architecture consists of two pathway (Path-I, Path-II), since it is a dual input CNN. Path-I takes input as binarized squared image patches and Path-II have skeletonized binary squared image patches as input. Both paths are part of feature extraction phase of our dual input CNN model. In the feature extraction phase, model generates the local feature maps after every convolution layer and then they down-sampled by pooling layer. Both path have same configuration of convolution and poling layers, as it consists of five convolution layers, with three max-pooling layers after first two and fifth convolution layers in order to increase the depth of network and two fully connected(dense) layers. The kernel/filter size of first convolution layers is 5 × 5 and 3 × 3 for rest convolution layers. The stride is 1

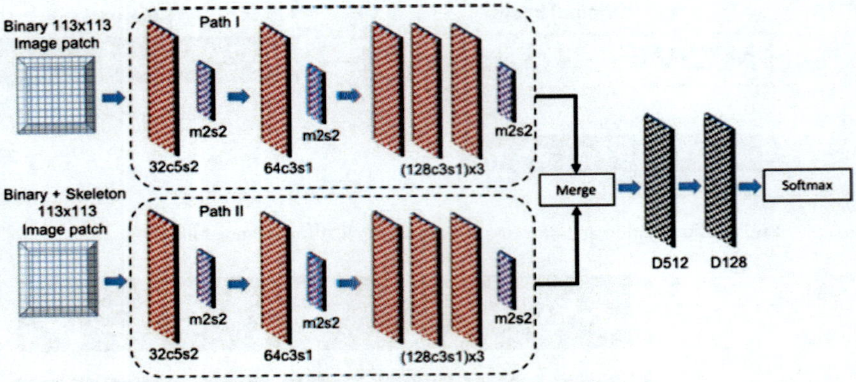

Fig. 6 Dual-Pathway CNN structure. In Fig. 32c5s2 notation means that convolution layer with 32 filter, 5 × 5 filter size and stride is 2. m2s2 means max pooling layer with filter 2 × 2 and stride = 2. D512 denotes that Dense layer with 512 input size

for all convolution layers and 2 for all pooling layers. The configuration of proposed model is shown in Fig. 6.

To keep the neurons activated in the network, We use ReLU (Rectified Linear Unit) function after every convolution and dense layer. If the neurons input values go below the threshold then they die or deactivate. ReLU returns 0 if it encounters any negative input value and for any positive value it returns that value back. So it can be written as:

$$ReLU(x) = \max(0, x). \tag{1}$$

In proposed model, We use max-pooling layers with kernel=3 × 3, stride = 2 after each convolution layer for down sampling the feature maps. There are total five consecutive convolution layer with max-pooling layer used in both pathways, with different input image patches. Both pathways generate several different local feature maps from binarized and skeletonized image patches. After the last pooling layer in both pathways, the down-sampled local feature maps, which are in 2D matrix shape converted into 1D vector array using flattening operation. After flattening, We merged 1D vector array of Path-I and Path-II into a single 1D vector array. This merged 1D vectors are passed as input to the next fully connected layers. As shown in Fig. 6, in the proposed model there are two hidden layers as fully connected layer with 512 and 128 nodes. These two dense layer makes the multi-layer perceptron network for training of the network. By the merged flattened 1D vector input the dense layers train the network. To reduce the over-fitting, We use dropout layer with 50% rate after each dense layer. After network training, last layer is Softmax, it is used for multi-class classification which predict the writer class of queried image document [12].

4 Experiments and Results

In this section, provide the dataset details, the experimental setup of our model and results. The IAM dataset is used to train and test our proposed model. This dataset consists handwritten English text from 657 different writers, using different pens. They provide whole document and segmented lines handwritten scanned images separately with PNG format. We used segmented line dataset which contain line images each one with one sentence. The experiments and model construction was implemented on google co-laboratory environment. Google co-laboratory is free and open-source platform and it provides 25–36 GB of virtual-RAM with 12GB GPU memory. We modified the IAM dataset according to our limited virtual resources. Sentence images contributed by each writer are divided into training set and testing set according to the ratio of (70:30). Dual-Pathway deep CNN achieves 92.7% accuracy with approx more than 120,000 image patches of 20 different writers.

TensorFlow and Keras libraries are used for implementation and training of the proposed model. Some morphological operations have done using OpenCV and skimage libraries during pre-processing stage. The model is trained with total number of 30–40 epochs with training and validation data in the ratio of (70:30). The visualization of accuracy and loss curves of training and validation data after model training is shown in Fig. 7.

We have also worked with size variations of squared image patch. Table 1 shows that decrements in patch size also affects the model accuracy. We also compared our proposed model performance to the other models and state-of-art methodologies as shown in Table 2. Dual-Pathway CNN is getting high accuracy rates than Sheng et al. [10] and Bulacu et al. [2].

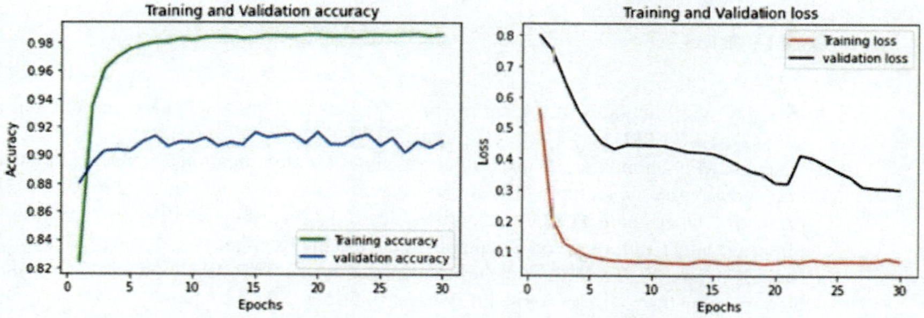

Fig. 7 Accuracy and loss curves for training and validation data

Table 1 Patch size comparison

Patch size	Accuracy %
113 × 113	92.7
82 × 82	89.8

Table 2 Comparison to other methods

Reference	Feature	Accuracy %
Sheng et al. [11]	Adaptive CNN	85.2
Sheng et al. [10]	FragNet CNN	85.1
Bulacu et al. [2]	Probability distribution function	89.9
Proposed	Dual-pathway CNN	**92.7**

5 Conclusion

This paper proposed a Dual-Pathway CNN network for writer recognition based on binary and skeleton image input patches feature map extraction. The proposed method applied on a modified IAM dataset, which consists of segmented line scanned images from different writers. To increase the accuracy of the model, we analyzed the features of image patches extracted by model. To attain higher accuracy of model, the image size and quality is also responsible. Therefore we perform preprocessing on image data before training of model. Validation data accuracy curve(blue) and loss curve (black) in Fig. 7, are moving upwards and downwards respectively with some fluctuations during training of the model. It shows that still we need some improvements in the model configuration. Our method can be extended to some other scripts. In future we plan to extend this method for other datasets. We will also work on patch extraction strategy and overfitting issues, since they also impacts the model performance.

References

1. Sreeraj, M., Idicula, S.M.: A survey on writer identification schemes. Int. J. Comput. Appl. **25**(12), 23–33 (2011)
2. Bulacu, M., Schomaker, L.: Text-independent writer identification and verification using textural and allographic features. IEEE Trans. Pattern Analysis Mach. Intell. **29**(4), 701–717 (2007)
3. Jain, R., Doermann, D.: Offline writer identification using K-adjacent segments. In: 2011 International Conference on Document Analysis and Recognition, Beijing, 2011, pp. 769–773
4. Wu, X., Tang, Y., Bu, W.: Offline text-independent writer identification based on scale invariant feature transform. IEEE Trans. Inf. Forensics Secur. **9**(3), 526–536
5. Jain, R., Doermann, D.: Combining local features for offline writer identification. In: 2014 14th International Conference on Frontiers in Handwriting Recognition, Heraklion, 2014, pp. 583–588
6. Xing, L., Qiao, Y.: Deepwriter: a multi-stream deep CNN for text-independent writer identification. In: 2016 15th International Conference on Frontiers in Handwriting Recognition (ICFHR), Shenzhen, Oct 2016, pp. 584–589
7. Christlein, V., Gropp, M., Fiel, S., Maier, A.K.: Unsupervised feature learning for writer identification and writer retrieval. In International Conference on Document Analysis and Recognition (ICDAR), vol. 1, pp. 991–997, 2017

8. Chen, S., Wang, Y., Lin, C.-T., Ding, W., Cao, Z.: Semi-supervised feature learning for improving writer identification (2018)
9. Krizhevsky, A., Sutskever, I., Hinton, G. E.: Imagenet classification with deep convolutional neural networks. In: Advances in Neural Information Processing (2012)
10. He, S., Schomaker, L.: FragNet: writer identification using deep fragment networks. IEEE Trans. Inf. Forensics Secur. **15**, 3013–3022 (2020)
11. He, S., Schomaker, L.: Deep adaptive learning for writer identification based on single handwritten word images. Pattern Recogn. **88**, 64 – 74 (2019)
12. Fiel, S., Sablatnig, R.: Writer identification and retrieval using a convolutional neural network. In: Computer Analysis of Images and Patterns, pp. 26–37. Springer
13. Rehman, Naz, S., Razzak, M.I., Ibrahim, H.A.: Automatic visual features for writer identification: a deep learning approach. IEEE Access **7**, 17149–17157 (2019)
14. Keglevic, M., Fiel, S., Sablatnig, R.: Learning Features for Writer Retrieval and Identification using Triplet CNNs. ICFHR, pp. 211–216, 2018
15. Adak, C., Chaudhuri, B.B., Blumenstein, M.: Impact of struck-out text on writer identification. In: Proceedings of 30th International Joint Conference on Neural Networks (IJCNN), pp. 1465–1471, 2017
16. Nguyen, H.T., Nguyen, C.T., Ino, T., Indurkhya, B., Nakagawa, M.: Text-independent writer identification using convolutional neural network. Pattern Recogn. Lett. **121**, 104–112 (2019)
17. Simonyan, K., Zisserman, A.: Very deep convolutional networks for large-scale image recognition. arXiv:1409.1556, 2014
18. He, K., Zhang, X., Ren, S., et al.: Deep residual learning for image recognition. arXiv:1512.03385, 2015
19. Ciresan, D., Meier, U., Schmidhuber, J.: Multi-column deep neural networks for image classification. In: 2012 IEEE Conference on Computer Vision and Pattern Recognition (CVPR), pp. 3642–3649. IEEE (2012)
20. Christlein, Maier, A.: Encoding CNN activations for writer recognition. In: Proceedings of 13th IAPR International Workshop Document Analysis System (DAS), pp. 169–174, Apr 2018

A Multi-Agent System for Industrial Simulators Design

Kenza Redjimi and Mohammed Redjimi

Abstract Industrial systems are of great importance in our daily life. They are, generally characterized by a huge number of interacting elements. Some of them are critical and cannot be stopped of functioning otherwise they will cause very significant financial damage to the plant. To observe the process of functioning of such systems is based on their modeling. This task is generally based on a fine division of sub-assemblies of components of less complexity. Each component can be modeled individually and then simulated. In this article, we present a top-down modeling method using a multi-agent systems approach. So, the overall system is considered as a set of interacting agents. The behavior of the global system emerges from the behavioral of its parts. This approach was applied to the liquefied natural gas treatment process (decarbonation) at a gas plant in Algeria. The results of the simulation can be used to train new recruits in this factory.

Keywords Multi-agent system (MAS) · Industrial system modeling and simulation · Agent/group/role (AGR) mModel · Aalaadin · Madkit

1 Introduction

Today, industrial systems employ advanced technologies for the development of complex industrial processes. The vast majority of these systems can be considered complex in the sense that they are composed of a large number of interacting elements. To develop such systems, to predict their behaviors before beginning to build them, or for educational training, these systems are modeled and undergo a set of tests.

The use of computer tools for modeling offers the advantage of building a computer model that is very faithful to the considered system, a set of its parts,

K. Redjimi (✉) · M. Redjimi
Faculty of Science, Department of Computer Science, LICUS Laboratory, 21000 Skikda, Algeria
e-mail: k.redjimi@univ-skikda.dz

M. Redjimi
e-mail: m.redjimi@univ-skikda.dz

L. Troiano et al. (eds.), *Advances in Deep Learning, Artificial Intelligence and Robotics*, Lecture Notes in Networks and Systems 249,
https://doi.org/10.1007/978-3-030-85365-5_13

then to judge the behavior of the obtained model by injecting it with a set of data as inputs. This makes it possible to handle, to observe, and to improve understanding of its behavior and allows avoiding any possible errors concerning its future evolutions. The shutdown of certain large so-called critical systems can cause material and financial damages. For these systems, the use of models representing them is of substantial help. In fact, instead of stopping the operation of all or part of a plant, it suffices to consider the models, which represent them. This is the case, for example, with industrial boilers, the shutdown of which can cause the shutdown of a whole part of the plant, which depends on it.

The notion of system is at the center of several fields and disciplines. A system is characterized by a set of communicating and interacting elements. A system receives inputs and delivers results. Various systems can be envisaged such as industrial systems, economic systems, political systems… A metric often used to characterize a system concerns its complexity.

The modeling of systems has been the subject of many studies and researches considering its important impact on the phenomena implemented. Fishwich's [1] considers modeling as a set of metaphors and analogies to better understand any phenomena. For Oussalah's [2], modeling is a transformation between a system and a model to give a simplified and observable representation of its structure and behavior.

The experimentation of the model can be considered as its simulation. Practically, a simulation is the execution phase of the model [3]. Given a set of pertinent data as input of the model, this phase consists of evolving the model and observing and analyzing the obtained results. So, works can be observed as modeling of actual cases in the laboratory.

Simulation of complex systems can take hours or even days and consume a lot of CPU time. This is why; modelers have recourse to methods of reduction of complexity. Among these methods is parallel simulation [4], decomposition of the phenomenon in several entities of fewer complexities [5], or the use of fast and parallel CPUs [6].

The search for high fidelity, precision, and adaptation requires that modeling and simulation data processing tools be chosen with great care. Their use in the field of modeling and simulation is very widespread, ensuring excellent qualities for which the improvements are continuous with reasonable costs.

Several modeling approaches, methods, and tools have been and continue to be used in the literature. Petri nets (PN) [7] are among the most efficient and widely used tools. Despite their 62 years of existence, they continue to be relevant today.

These networks, despite their simplicity, make it possible to represent the dynamics of systems. Several variations and extensions of these networks have been developed such as Colored PN (CPN) and Timed PN [8, 9].

The UML state chart diagrams are other powerful semi-formal tools that can represent the behavior of systems [10]. At a higher level, multi-agent systems (MAS) are modeling tools that offer the advantage of a fine decomposition of a real complex system into a set of its components. The behavior of the global system emerges from that of the behaviors of its entities. The notion of agent designates an entity whose functioning is autonomous and which is capable of carrying out a set of tasks [11, 12]. These systems are used more and more in computer systems because of their flexibility, their simplicity, their effectiveness, and their possibility of extension. Moreover, a human observer can easily understand the description given by MAS.

The rest of this paper is organized as follows: Sect. 2 presents the multi-agent systems-based approach. Section 3 is dedicated to the proposed approach. Section 4 introduces a use case of industrial system simulation, and finally, Sect. 5 concludes this paper.

2 The Multi-Agent System (MAS)

Multi-agent systems (MAS) provide high-level tools for modeling distributed systems. Their application has become, nowadays, very widespread and they have been applied with success in various domains especially those based on artificial intelligence. A set of autonomous agents cooperate and collaborate in the execution of a common objective. From the behavior of all these agents emerges the overall behavior of the system [11–20].

2.1 *Agent Definition*

There is no common agreement about a definition of the term agent. In general, an agent is an entity (hardware or software), which is different from an ordinary program in the sense that it is endowed with autonomy for functioning in environments shared by other agents [12, 16]. An agent is able to evolve in an environment through the acquisition of information emitted by the latter and by acting on it. The architecture of an agent varies from a simple structure formed of production rules often linked to very complex structures where the agent can be endowed with advanced mechanisms of cognition and reasoning referring to it as an associative and social character [21].

3 The Proposed Approach

The work concerned by this present theme is related to the development of industrial process simulators in the hydrocarbon sector. This involves providing educational support to technical staff so that they can understand the different stages of the process.

The proposed approach is a top down approach, which starts, therefore, from the overall process and then divides it up according to the fundamental units (see Fig. 1) after having identified them.

The process cutting procedure is carried out as follows:

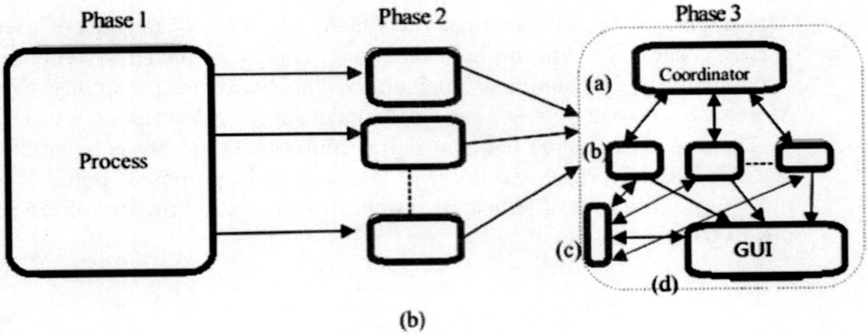

Fig. 1 Process breakdown

1. The process is studied in detail. The principle of its operation is clearly established.
2. The different units making up the process are identified according to the detailed technical diagram of the process (b).
3. Each unit is isolated to be individually modeled and the obtained units are assembled in a global diagram.

Three additional control and communication units are added to the model:

a. A coordinator whose role is to control and manage the execution of the whole system.
b. An input interface through which the user can act on the system.
c. An interface for visualizing the progress of the process according to the inputs sent by the user.

3.1 Tools for the System Design

For the design of this kind of simulators, a solution using multi-agent systems is very appropriate. An agent represents each unit. The same goes for the coordinator, the input interface as well as the display interface.

Nowadays, there are several multi-agent systems-based approaches well suited to the modeling and simulation of systems. Likewise, a wide variety of multi-agent platforms is available. Among the existing methodologies, the AGR (Agent-Group-Role) model gives very good results in terms of modularization of systems.

In the AGR model, an agent can play one or more roles in one or more groups. A group is made up of at least one agent. An agent who does not belong to a group can request to join this group. Several agents can play the same role. An agent can communicate with another only if the two agents are in the same group. The interactions between agents are founded on the messages exchange principle and allow coordination, collaboration and synchronization between agents.

The AALAADIN meta-model is an organizational model for the modeling of multi-agent systems based on AGR concepts. A set of conceptual tools for the implementation of applications based on multi-agent systems are provided by AALAADIN. AALAADIN was developed at the computer science, robotics and micro-electronics laboratory, Montpellier (LIRMM) [22–24].

3.2 The MADKIT Platform

The Multi Agent Development Kit platform (MadKit) is a free software for the design and the simulation of Multi-Agent Systems (MAS) [25–29]. MadKit is a set of packages of Java classes. It implements a reduced size agent kernel, various libraries of messages, probes and agents. In addition, Madkit includes a useful graphical development environment and many other agent handling systems. Madkit implements different tools for the management of agents, groups and roles according to the AGR model. This platform is based on the AALAADIN meta-model.

4 Use Case

Pure natural gas (NG) is a fuel unfit for consumption due to the presence of impurities, particularly water and carbon dioxide (H_2O and CO_2), which reduces its calorific value. Liquefaction of NG requires cooling to about −165 °C. Knowing that the freezing temperature of the water is at 0 °C and that of the CO_2 at −20 °C, this will clog the pipes of the machines during the liquefaction process. It is for these reasons that NG must be treated (decarbonized and dehydrated) before the liquefaction operation. Decarbonation is the extraction of carbon dioxide CO_2 from natural gas and dehydration consists of extracting water from it. There are several industrial ways to execute the decarbonation. [30] Presents the process adopted in this work.

4.1 The Natural Gas Decarbonation Process

This use case are two operations for the decarbonation process: (i) the washing operation, which is intended to extract the CO_2 from the NG with an M.E.A ($C_2H_5ONH_2$), and the regeneration phase, where the results from the washing operation enters a regeneration tower so that they release CO_2 and then restore the MEA solution. Figures 2 and 3 illustrate the dedicated equipment.

According to the breakdown system phases presented Fig. 1, at first and after studying and analyzing the whole system as presented in Figs. 3 and 4, the following four points are highlighted:

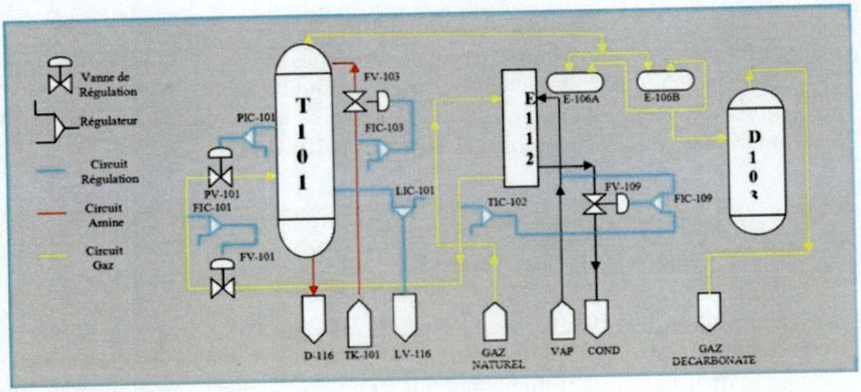

Fig. 2 Technical drawing of the washing process

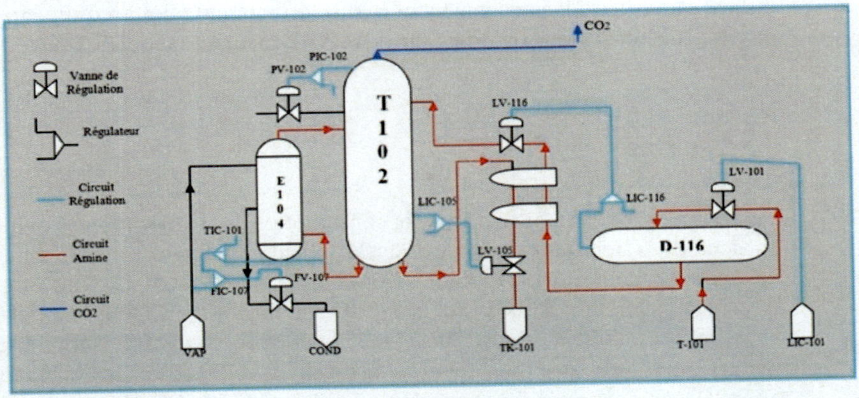

Fig. 3 Technical drawing of the regeneration process

- **The system ordered**: it is the decarbonation process. It contains a set of organs (T-101, T-102, D-116, E-112, E-104 …) which carry out the various chemical reactions to achieve the decarbonation of the Natural Gas.
- **The control system**: this is the distributed control system for the natural gas decarbonation process. It is formed by a set of autonomous, modular and dis- tributed instruments. They are intended for the control and regulation of the process.

These instruments are divided into two sections:

- **The washing section**: it groups the regulation valves (FV-109, FV-103, FV-101, PV-101), which are regulated by a control system, which contains the different regulation loops.

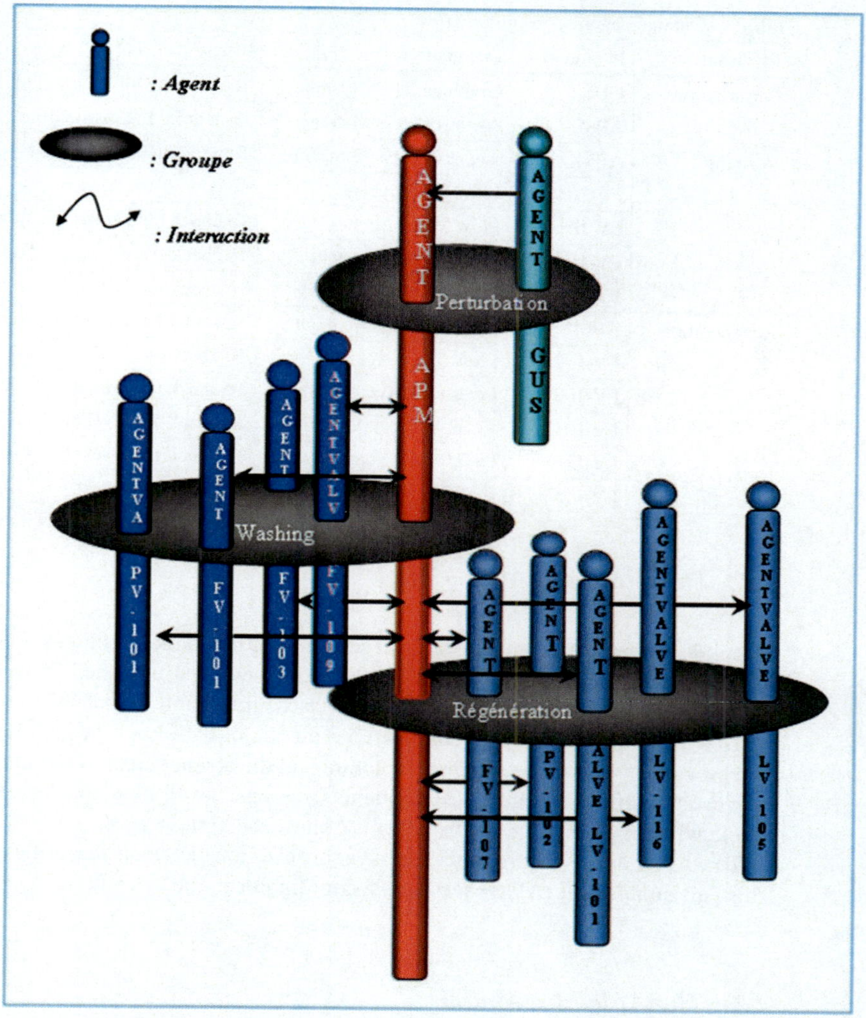

Fig. 4 The AALAADIN model of the simulator

- **The regeneration section**: also contains a set of control valves (FV-107, PV-102, LV-116, LV-105, LV-101), which are regulated by the same control system. To carry out this adjustment the control system is based on the set of instructions.

Table 1 Obtained agents, groups and roles

Id_Group	Id_Agent	Designation	Role
Perturbation	GUS	Graphical User System	To interact with the system
	APM	Agent Process Manager	To acquire the Instructions
Washing	APM	Agent Process Manager	To regulate the washing group
	FV-103	Flow Valve Agent	To control the Amine flow
	FV-101	Flow Valve Agent	To check the natural gas flow
	PV-101	Pressure Valve Agent	To check the pressure of T-101
	FV-109	Flow Valve Agent	To check the steam flow of E-102
Regeneration	APM	Agent Process Manager	To set the regeneration group
	FV-107	Flow Valve Agent	To check steam flow E-104
	PV-102	Pressure Valve Agent	To check the pressure T-102
	LV-101	Level Valve Agent	To check the level of T-101
	LV-116	Level Valve Agent	To check the level of D-116
	LV-105	Level Valve Agent	To check the level of T-102

4.2 The Proposed AGR Model

To build an interactive simulator for the above-mentioned industrial gas process, the three following groups as depicted Table 1 have been distinguished. Note that this point corresponds to the phase 2 of the proposed simulation approach:

This decomposition was obtained after a meticulous study of the NG decarbonation process as well as the material structure and the arrangement of the interactions between the different material components (pressure, level, flow control, etc.). This article will not go into the precise details of how the system works.

The APM agent ensures the time synchronization between the different units. This particular agent (APM) serves as a coordinator in the system.

4.3 The Aalaadin Model

The entire model with links between all the components (agents) of the system is depicted by the Fig. 4 below. This diagram does not describe the roles of each agent. This step corresponds to the phase 3 of the proposed demarche.

4.4 The System Implementation

The simulator was developed by using the Mad-Kit agent platform based on AALAADIN model. The two main interfaces of the project are presented in Figs. 5

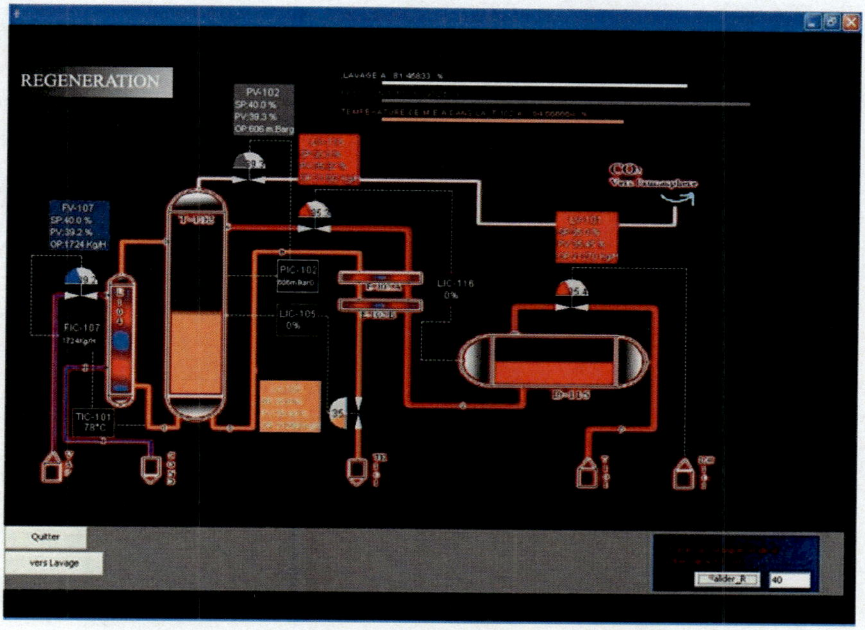

Fig. 5 The interface of the washing section simulator

and 6. These two interfaces concern the washing section and the interface of the regeneration section. They allow the user to drive the washing and the regeneration operations and to supply the system by data inputs.

4.5 *Discussion*

This work aims to implement the MAS mechanisms of cooperation, collaboration and synchronization between agents. These tools are used to design and to realize industrial simulators. The use case presented and implemented here, which can be very useful as a pedagogical support, consolidates and validates the proposed approach.

5 Conclusion

The concerns of this work are to introduce some notions of the design of industrial simulators. This work responds to a real need of providing a tool allowing young beginners in the company to understand the evolution of an industrial process. In this case, it is the process of liquefied natural gas decarbonation. The equipment allowing

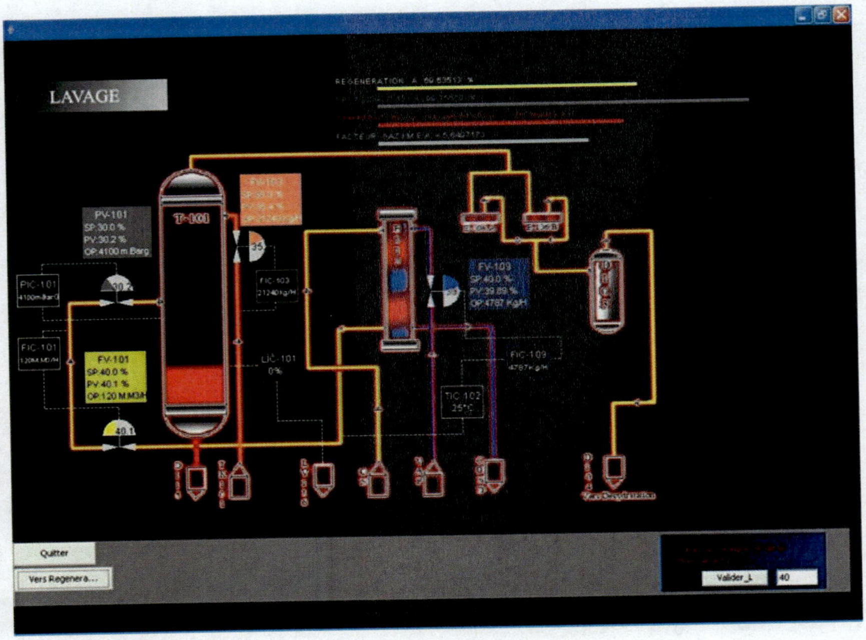

Fig. 6 The interface of the regeneration section simulator

this procedure is modeled using agents. The coordination and cooperation between these agents allow the exchange of information and the evolution of the process. This simulator is currently operational and its use allows significant financial and material gains.

All the components of the system are modeled using the agent paradigm. The interaction between these agents ensures the evolving of the system and the global behavior emerges from the local agent's execution. The use of consistent and coherent tools (for example, AALAADIN and MADKIT) is of great interest in obtaining efficient software products, as is the case here. In the future work, we project to extend this approach to distributed systems modeling.

References

1. Fishwick, P.: Simulation Model Design and Execution. Building Digital Worlds (1995)
2. Oussalah, C.: Modèles hiérarchisés multi-vues pour le support de raisonnement dans les domaines techniques. Technical report (1988)
3. Fishwick, P., Paul. A.: Computer simulation: growth through extension. Trans. Soc. Comput. Simulat. Int. **14**(1), 13–23 (1997)
4. Mubarak, M., Carothers, C.D., Ross, R.B., Carns, P.: Enabling parallel simulation of large-scale HPC network systems. IEEE Trans. Parallel Distrib. Syst. **28**(1), 87–100 (2016). https://

doi.org/10.1109/TPDS.2016.2543725
5. Wainer, G., Liu, Q., Jafer, S.: Parallel simulation of DEVS and Cell-DEVS models in PCD++. Discrete-Event Modeling and Simulation **20115630**, 223–270 (2016)
6. Martınez-del-Amor, M.A., Riscos-Núnez, A., Pérez-Jiménez, M.J.: A survey of parallel simulation of P systems with GPUs. Bulletin Int. MC Soc. 55–68 (2017)
7. Juranic, J., Pavkovic, N., Naumann, T., Marjanovic, D.: Modelling the design parameters dynamics with Petri nets. In: DS 87–2 Proceedings of the 21st International Conference on Engineering Design (ICED 17) Vol 2: Design Processes, Design Organisation and Management, 091–100. Vancouver, Canada (2017)
8. Jensen, K., Kristensen, L.M.: Coloured Petri Nets, Springer, Berlin Heidelberg (2009). https://doi.org/10.1007/b95112
9. Jensen, K., Kristensen, L.M., Wells, L.: Coloured Petri Nets and CPN Tools for mod- elling and validation of concurrent systems. Int. J. Softw. Tools Technol. Transfer **9**(3–4), 213–254 (2007). https://doi.org/10.1007/s10009-007-0038-x
10. Sunitha, E.V., Samuel, P.: Automatic code generation from UML state chart dia- grams. IEEE Access **7**, 8591–8608 (2019). https://doi.org/10.1109/ACCESS.2018.2890791
11. Mess, M., Guerrits, B.: Multi-agent Systems, Lecture Notes in Logistics, 611–636. (2018). https://doi.org/10.1007/978-3-319-92447-2-27
12. Ferber, J.: Les systèmes multi-agents: versune intelligence collective. Informatique, intelligence Artificielle, Intereditions Paris (1995)
13. Chen, F., Ren, W.: On the control of multi-agent systems: a survey Foundat Trends Syst Control **6**(4), 339–499 (2019)
14. Weiss G. (ed): Multiagent Systems—A Modern Approach to Distribute Artificial Intelligence. MIT Press, Cambridge, MA (1999)
15. Wang, B., Wang, J., Zhang, B., Li, X.: Global cooperative control framework for multia- gent systems subject to actuator saturation with industrial applications. IEEE Trans. Syst. Man Cybernet. Syst. **47**(7), 1270–1283 (2016)
16. Ferber, J.: Multi-Agent Systems. Addison-Wesley, An Introduction to Distributed Artificial Intelligence (1999)
17. Shoham, Y.: Agent-oriented programming. Artific.l Iintell. **60**(1), 51–92 (1993)
18. Franklin, S., Gasser, A.: Is it an agent, or just a program? A taxonomy for autonomous agents. In: Muller, W., Jennings (eds). Intelligent Agents III. Agent Theories, Architectures, and Languages, 21–35. Springer Verlag (1997)
19. Wooldridge, M., Jennings, N.R.: Agent theories, architectures, and languages. In: Wooldridge, and Jennings, (eds). Intelligent Agents, 1–22. Springer Verlag (1995)
20. Russell, S., Norvig, P.: Artificial Intelligence: A Modern Approach. Pearson (2010)
21. Ye, P., Wang. S., Wang, F.: A general cognitive architecture for agent-based modeling in artificial societies. IEEE Trans. Comput. Soc. Syst. **5**(1), 176–185 (2018). https://doi.org/10.1109/TCSS.2017.2777602
22. Ferber, J., Gutknecht, O., Michel, F.: From agents to organizations, an organizational view of multi-agent systems. Agent Oriented Software Engineering (AOSE) IV, P, 2935, 214–230 (2004)
23. Souidi, M.E.H., Songhao, P., Guo, L., Lin, C.: Multi-agent cooperation pursuit based on an extension of AALAADIN organizational model. J. Exp. Theor. Artif. Intell. **28**(6), 1075–1088 (2016). https://doi.org/10.1080/0952813X.2015.1056241
24. Russell, S.J.: Rationality and intelligence. Artif. Intell. **94**, 57–77 (1997)
25. Ferber, J., Gutknecht, O.: Aalaadin: a meta-model for the analysis and design of organizations in multi-agent systems. In: Demazeau Y (ed). ICMAS (International Conference on Multi-Agent Systems), 128–135. IEEE Press, Paris (1998)
26. Chebout, M.S., Mokhati, F., Badri, M., Babahenini, M.C.: Monitoring open multi-agent systems: AAn aspect-oriented programming based approach. Multiagent Grid Syst. **15**(2), 155–177 (2019). https://doi.org/10.3233/MGS-190307
27. http://mansour.saber.free.fr/termadkit/site/madkit/doc/devguide/devguide.html. Accessed on 24 Sept 2020

28. Gutknecht, O., Ferber, J.: The MadKit agent platform architecture. Laboratoire d'Informatique de Robotique et de Microélectronique de Montpellier (2000).
29. http://www.madkit.net/madkit/. Accessed on 24 Sept 2020
30. https://patents.google.com/patent/WO2014199036A2/en. Accessed on 24 Sept 2020

Comparison of Current Convolutional Neural Network Architectures for Classification of Damaged and Undamaged Cars

Yavuz Ünal, Şaban Öztürk, Muhammed Nuri Dudak, and Mehmet Ekici

Abstract Deep learning algorithms, one of the most popular topics of recent years, are frequently encountered in many areas. One of these areas is traffic. Many applications such as vehicle identification and damage detection using images obtained by various imaging methods in traffic are successfully carried out by deep learning algorithms. In this study, damaged and undamaged car images are detected using current convolutional neural network architectures. In this study, the publicly published data set named damaged car dataset (DCD)-2 was used. This data set includes 80 normal and 80 accident vehicle images. The low number of images in the preliminary analyzes made it cause us to increase the data (image augmentation). With the increase in data, the number of images was increased to 400 damaged vehicles and 400 solid vehicles. VGG-16, VGG-19, Inception V3, NasNetLarge, DenseNet201 were applied from the transfer learning algorithm to the data incremented dataset. At the end of the application, the classification accuracy was 84.11% with VGG-16, 81.96% with VGG-19, 94.11% with Inceptionv3, 98.39% with NasnetLarge and 97.14% with Densenet201.

Keywords Deep learning · Car crash · Transfer learning

1 Introduction

Deep learning is one of the sub-branches of machine learning. In the process that has passed since the discovery of machine learning, the need and usage area of artificial intelligence have gradually increased, the most used artificial intelligence algorithms and deep learning architectures have emerged today. With the development of deep learning architectures, many deep learning approaches have emerged to solve artificial intelligence problems. It produces solutions that can be used in many areas such as industry, medicine, robotics, image processing, computer vision, object detection, sound processing-recognition, translation, future prediction, financial [1].

Y. Ünal (✉) · Ş. Öztürk · M. N. Dudak · M. Ekici
Amasya University, Amasya, Turkey
e-mail: yavuz.unal@amasya.edu.tr

L. Troiano et al. (eds.), *Advances in Deep Learning, Artificial Intelligence and Robotics*, Lecture Notes in Networks and Systems 249
https://doi.org/10.1007/978-3-030-85365-5_14

Convolutional Neural Network (CNN) is the most effective deep learning architecture used to solve image processing problems. The calculation cost of CNN training is high due to the excessive number of layers and nerve cells. Also, it may not always be possible to have a problem-specific training set. Weight transfer from the layers of a CNN trained with large and comprehensive training data set with proven classification success is the commonly used method [2].

When similar studies with the same dataset are examined; Ravindran et al. [5] proposed a new approach system that detects automatic road accidents using (DCD)-1 and (DCD)-2 datasets in their study. Accordingly, they used five different support vector machines using Histogram of gradients (HOG) and Gray level co-occurrence matrix (GLCM) features. They achieved 81.83% classification accuracy for the DCD-1 data set and 64.37% for the DCD-2 dataset. In the study conducted by Patil et al. [3], they classified 7 different types of accident vehicles by deep learning on their own accident vehicle data sets. Because of the scarcity of images they had, they first performed data augmentation and then classified them with Inception, Alexnet, VGG-19, VGG-16, Resnet transfer learning algorithms. They found the highest classification accuracy with 89.5%. Dhieb et al. In their study [4], detected the damaged parts of the crashed vehicles through deep learning. First of all, they increased the images of the damaged vehicles with data augmentation. Then they analyzed it with inception-Resnet-V2.

In this study, it is aimed to determine a classifier that can automatically separate accident-free and accident-free vehicles. For this reason, CNN architectures, which are the most powerful image processing methods in the literature, are used. The Classification process is carried out using VGG-16, VGG-19, Inception V3, NasNetLarge and DenseNet201 networks. In this process, these networks, which are pre-trained, are lightly trained with the dataset. After the fine-tuning process, the performances of the current CNN architectures are compared. The results obtained show the usability of the original CNN architectures for this process.

This study is organized as follows. In Part II, the theoretical infrastructure of this study, the database and methods used are explained. The results obtained in Chapter III are given. In Section IV, results and recommendations are given.

2 Materials and Methods

The technical information required for the classification of cars in the DCD-2 dataset and the details of the data are presented in this section. The deep learning methods used are deep feature extraction and transfer learning. In the study, experimental studies were carried out with pre-trained VGG-16, VGG-19, Inception V3, NasNet-Large, DenseNet201 models. The theoretical explanation of the CNN models used and the proposed method are given respectively. Figure 1 shows the operation of the proposed framework. Accordingly, image augmentation is applied to the images in the data first. In the Augmentation process, 90-degree rotation, 270-degree rotation, mirroring, noise addition procedures are applied. With the augmented dataset

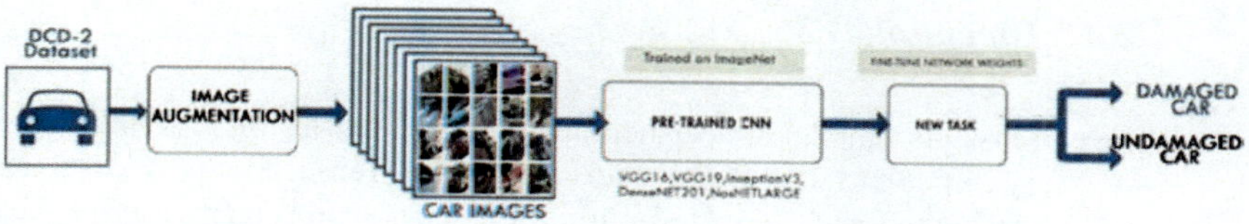

Fig. 1 Flow chart of the study

obtained, CNN architectures are trained sequentially. Then, these trained CNN architectures are tested to measure crash detection performance. As a result, the performance of each CNN architecture is compared.

2.1 *Transfer Learning*

Transfer learning is a research problem in machine learning that focuses on storing the information obtained in solving a problem and then applying it to a different problem [6].

2.2 *Vgg-16*

The VGG-16 deep learning algorithm is a network of 13 convolutions 3 fully connected layers used by Oxford University Visual Geometry Group to achieve better results in the ILSVRC2014 competition. There are total of 41 layers including maxpool, fullconnectedlayer, relulayer, dropoutlayer and softmaxlayer layers. The image to be included in the input layer is 224 × 224 × 3. The last layer is the classification layer. It is a deep learning algorithm that has achieved 89% accuracy in the ImageNet database [7].

2.3 *Vgg-19*

The VGG-19 deep learning algorithm is a network of 16 convolutions 3 fully connected layers used by Oxford University Visual Geometry Group to achieve better results in the ILSVRC2014 competition. There are total of 41 layers including maxpool, fullconnectedlayer, relulayer, dropoutlayer and softmaxlayer layers. The image to be included in the input layer is 224 × 224 × 3. The last layer is the classification layer. It is a deep learning algorithm that has achieved 89% accuracy in the ImageNet database [7].

2.4 Inception-V3

Inception-v3 is a model of deep neural networks. With a simple configured network, it is very difficult to train it and takes a long time. Tensorflow provides a tutorial for us to retrain the Last Tier of Inception for new categories using transfer learning. It uses the transfer learning method, which preserves the parameters of the previous layer and removes the last layer of the Inception-v3 model, then retrains the last layer [8].

2.5 NasnetLarge

NASNet-Large is a CNN network trained on over one million images using the ImageNet database. Network, images keyboard, mouse, pen, etc. Categorize as 1000 different objects. The image input size of the network is 331 × 331 pixels [9].

2.6 DenseNet201

The DenseNet CVPR2017 proposed by Huang et al. [9] made a name for itself by receiving the best article award. The Densenet-201 model is also one of the DenseNet group models designed for image classification. The main difference between the Densenet-121 model is the size and accuracy of the model [10].

2.7 Dataset Description

In this study, the public dataset damaged cars dataset-2 (DCD-2) was used [11]. There are two classes in this dataset. This dataset consists of 160 images in total, including 80 images of damaged and 80 solid vehicles. Some of the images in the dataset are in Fig. 2.

3 Results

CNN models and classification results used in this study are given in Table-2. Experimental results Calculations were performed on a virtual server with 2 Core Intel Xeon CPU, 12 GB ram and 16 GB Nvidia Tesla T4 GPU allocated by Colab through the Keras library in the Colab environment. Networks to be used for Transfer Learning were chosen among the networks trained with ImageNet. The networks used are

Fig. 2 Example images from the training and test dataset used

VGG-16, VGG-19, Inception V3, NasNetLarge, DenseNet201 and the parameters of these networks are given in Table 1. Networks in Table 1 are quite deep and the number of labeled samples required for training deep networks is high. The number of tagged samples in the dataset is not enough to train any network. For this reason, the number has been increased with augmentation. For this reason, the performance values of this network are not presented in Table 2.

Table 1 CNN networks used

Model	Size	TOP-1 ACC	Parameters	Depth
VGG-16	528 MB	0,713	138,357,544	23
VGG-19	549 MB	0,713	143,667,240	26
InceptionV3	92 MB	0,779	23,851,784	159
DenseNet201	80 MB	0,773	20,242,984	201
NasNetLarge	343 MB	0,825	88,949,818	-

Table 2 Training accuracy of 5 models

Model	Accuracy (%)	Time (sec)
VGG-16	84.11	1600
VGG-19	81.96	1600
InceptionV3	94.11	1400
NasNetLarge	98.39	4200
DenseNet201	97.14	2000

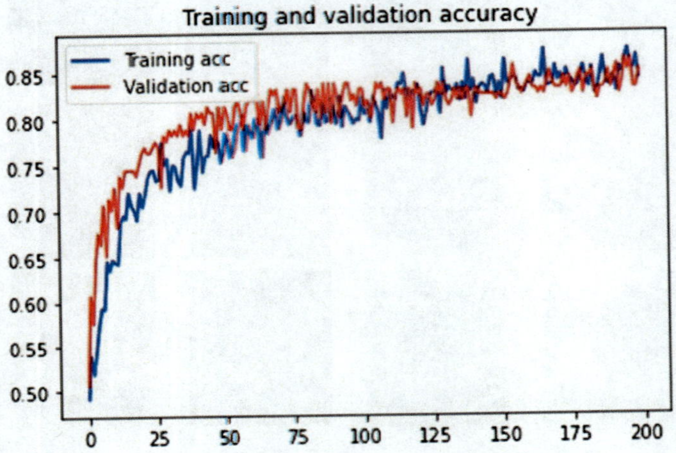

Fig. 3 VGG-16 training curve

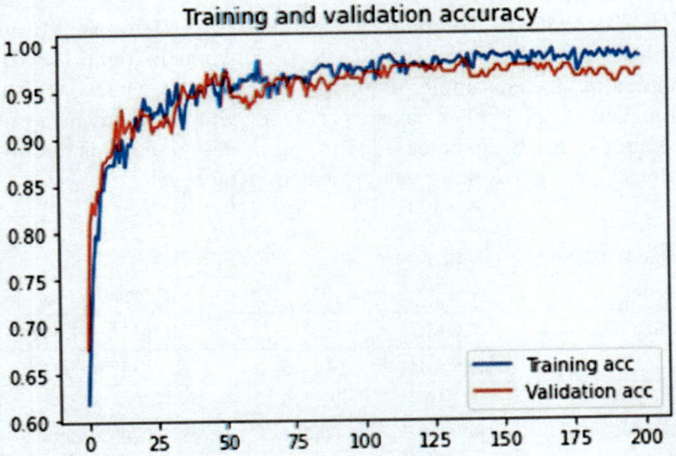

Fig. 4 VGG-19 training curve

Figure 3 shows VGG-16 training and validation performance. Fig. 4 VGG-19 training and validation performance, Fig. 5 Inception V3 training and validation performance, Fig. 6 NasNetLarge Training Curve training and validation performance, Fig. 7 DenseNet201 Training Curve training and validation performance.

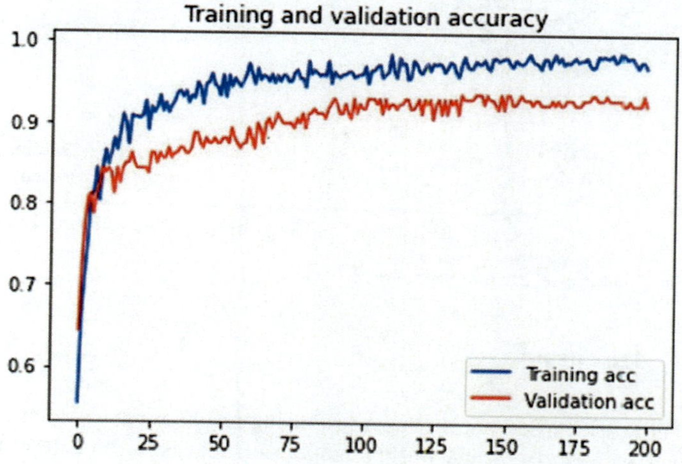

Fig. 5 InceptionV3 training curve

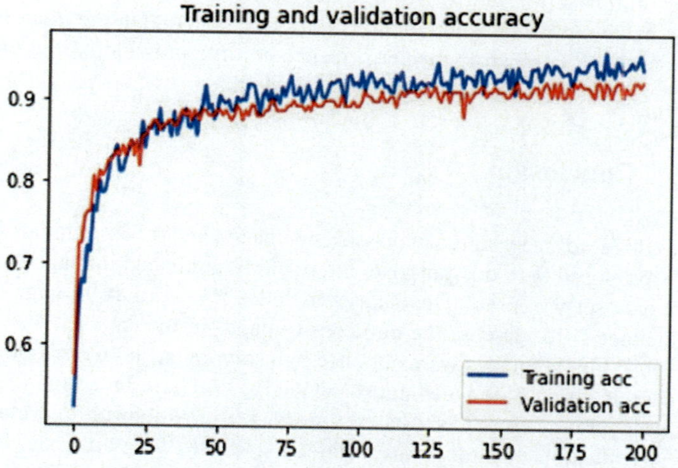

Fig. 6 NasNetLarge training curve

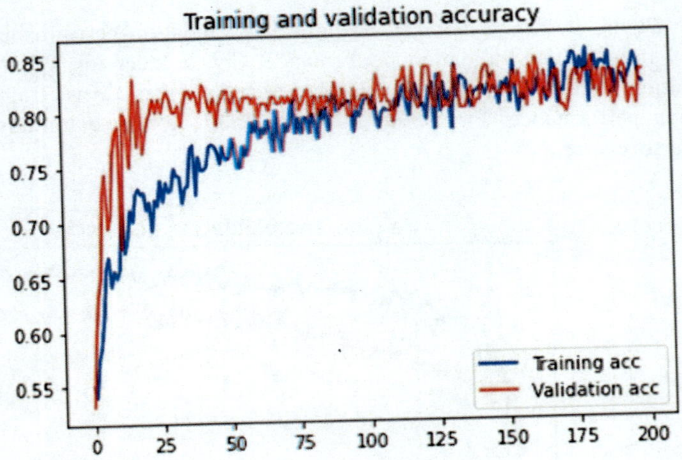

Fig. 7 DenseNet201 training curve

4 Discussion

The work carried out produces promising results for the detection of accident-free and accident-free cars. But the number of images in the dataset is very small. By increasing the number of images in the dataset, studies that produce clearer performance results can be presented. Because the images reproduced with augmentation are almost 4 times the original data. For this reason, it does not make sense, for now, to make performance predictions for a live system. On the other hand, this data set is ideal for comparing the performance of artificially increasing the number of data.

5 Conclusion

In this study, a system that classifies damaged and robust vehicles with deep learning is proposed. For this purpose, the publicly published damage car dataset (DCD) - 2 data set was used. This dataset includes a total of 160 images with 80 intact 80 damaged. To increase the number of images in the data set, 400 damaged and 400 solid vehicle images were obtained by using image augmentation. We tried multiple deep learning-based techniques. VGG-16, VGG-19, InceptionV3, Densenet201 and NasNetLarge from CNN-based transfer learning algorithms were applied. The best classification accuracy was obtained with 98.39% with NasNetLarge. In future studies, it is anticipated that studies can be made to determine the type of damaged parts from the images of the accident.

References

1. Doğan, F., Türkoğlu, İ: Derin öğrenme modelleri ve uygulama alanlarına ilişkin bir derleme. Dicle Univ. J. Eng. **10**(2), 409–445 (2019)
2. Fırıldak, K., Talu, M.F.: Investigation of transfer learning approaches used convolution neural networks. Anatolian J. Comput. Sci. **4**(2), 88–95 (2019)
3. Patil, K., Kulkarni, M., Karande, S.: Deep learning based damage classification. In: 2017 16th IEEE International Conference on Machine Learning and Applications (ICMLA), pp. 50–54 (2017)
4. Dhieb, N., Ghazzai, H., Besbes, H., Massoud, Y.: A very deep transfer learning model for vehicle damage detection and localization, microelectronics (ICM) 2019 31st International Conference on, pp. 158–161, (2019).
5. Ravindran, V., Viswanathan, L., Rangaswamy, S.: A novel approach to automatic road-accident detecting using machine vision technique. Int. J. Adv. Comput. Sci. Appl. (IJACSA) **7**(11), 235–242 (2016)
6. Pan, S.J., Yang, Q.: A survey on transfer learning. IEEE Trans. Knowl. Data Eng. **22**(10), 1345–1359 (2010)
7. Chen, C.L., Mahjoubfar, A., Tai, L.C., Blaby, I.K., Huang, A., Niazi, K.R., Jalali, B.: Deep learning in label-free cell classification. Sci. Rep. **6**, 21471 (2016)
8. Qian, Y., et al.: Fresh tea leaves classification using inception-V3. In: 2019 IEEE 2nd International Conference on Information Communication and Signal Processing (ICICSP), pp. 415–419, Weihai, China (2019)
9. Kaiming, H., Xiangyu, Z., Shaoqin, R., Jian, S.: Deep residual learning for image recognition. In: Proceedings of the IEEE International Conference on Computer Vision (2015)
10. Huang, G., Liu, Z., van der Maaten, L., Weinberger, K.: Densely connected convolutional networks. In: Proceedings of CVPR, pp. 2261–2269. Honolulu, HI, USA (2017)
11. DCD, [Online]. Available: https://github.com/vaishnavi29/DCD. Accessed 05 Sept 2020

Intelligent Agents for Multi-user Preference Elicitation

Ons Nefla, Iméne Brigui, Meltem Ozturk, and Paolo Viappiani

Abstract Preference elicitation is a central problem in AI including several approaches such as pairwise preference learning. Preferences are important in individual as well as in collective decision making such as social choice or negotiation. In this context, intelligent agents help modelling human behaviour and learn preferences in an incremental and autonomous way.In this article we aim to automate the preference elicitation process based on a multiagent system. Our multiagent architecture is made up of a set of autonomous user agents that communicate with the associated human users to learn his (her) preferences. The elicitation process uses the incremental MR-Sort method for ordinal classification problems. The proposed approach consists on an incremental process where questions of type "where do you assign alternative x?" are asked one by one. After each answer a new decision model is processed and the "best" question to be asked is found using different agent strategies.Different preferences of several agents are aggregated using voting strategies. A cost effective framework is also conceived to compute the most effective elicitation in order to improve the group decision making.

Keywords Multiagent systems · Preference elicitation · Ordinal classification · MR-Sort

---

O. Nefla · I. Brigui (✉)
Emlyon business school, Écully, France
e-mail: brigui-chtioui@em-lyon.com

O. Nefla · M. Ozturk
Paris Dauphine University, Paris, France

P. Viappiani
CNRS, LIP6 UMR 7606,Paris, France

L. Troiano et al. (eds.), *Advances in Deep Learning, Artificial Intelligence and Robotics*, Lecture Notes in Networks and Systems 249,
https://doi.org/10.1007/978-3-030-85365-5_15

1 Introduction

Preference elicitation addresses the problem of developing decision support systems able to assist users by modeling their preferences accurately and/or by generating recommendations.

Several Preference elicitation methods [1–4] allow to elicit incrementally a particular user's preferences by asking questions carefully chosen. These methods includes Bayesian approach [5] and Setwise max-margin optimisation [3]. Bayesian approach is proceeding by asking sequentially informative questions that improves the decision maker's expected utility. It derives relevant recommendation by asking relatively few questions to the decision maker. However, it is computationally expensive. Setwise max-margin optimisation is also proposed to elicitate incrementally user's preferences. Such a method has several purposes:it produces a set of "diverse" alternatives used to ask informative questions to the user; it encourage the sparse of utility functions in terms of weight combinations; it optimizes the recommendation by taking into account the collected information and it deals with inconsistencies with respect to decision maker responses. The recommendation is made by solving a constrained optimization problem.

These works focus on acquiring the preferences of a single user. However, in several real domains such as electronic commerce, management and industry, interaction occurs with several users at the same time. In addition, these users may provide little information about their preferences. Hence it is important to exploit the available preferences of similar users. Several works consider the problem of eliciting the preferences of a group of users at the same time [4, 6, 7]. However they didn't make use multiagent theory.

Multiagent theory [8] has been used in several research works to infer preference models mainly in recommender systems [9], clustering [10]. These methods allow to deduce user preferences based on their interactions and on users similarities.

In this paper, we consider the problem of eliciting the preferences of several decision-makers at the same time based on multiagent theory. First, we propose a distributed generic multiagent architecture. We will then focus on the agent's strategies for eliciting human preference incrementally using our multicriteria method : Incremental MR-Sort [11].

The paper is structured as follows. After a brief background in Sect. 2, we present our multiagent architecture in Sect. 3 and the user's agent elicitation module in Sect. 4. We then provide some final comments and perspectives in Sect. 6.

2 Background

2.1 Preliminaries

Let $X = \{1, \dots, m\}$ be a finite set of objects. A *binary relation* on X is denoted by $\succeq$, $x \succeq y$ meaning that x *is at least as good as* y. We denote by $\succ$ (resp. by $\sim$) the asymmetric (resp. the symmetric) part of $\succeq$: $x \succ y \iff x \succeq y$ and $y \not\succeq x$ (resp. $x \sim y \iff x \succeq y$ and $y \succeq x$). The relation $\succ$ represents the *strict preference* and the relation $\sim$ the *indifference*.

Objects of X are evaluated with respect to a *set of criteria* denoted by $Cr = \{cr_1, cr_2, \dots, cr_n\}$. We assume, without loss of generality, that the performance of each criterion is to be maximized. The *evaluation of* x is as a vector $(x_1, x_2, \dots, x_n)$; the evaluations of different criteria may be defined on different scales. Each criterion cr_i is associated with a *weight* w_i ($w_i \in W$), that intuitively represents its importance.

The *set of categories* is denoted by $C = \{C_1, C_2, \dots, C_p\}$. Categories are ordinal, C_1 being the worst one and C_p the best one. Each category C_h is characterized by two "fictitious" objects $b^h = (b^h_1, ..., b^h_n)$ and $b^{h+1} = (b^{h+1}_1, ..., b^{h+1}_n)$; these are called the *limit profiles of* C_h and denote by $B = \{b^1, b^2 \dots, b^{p+1}\}$ the set of such limit profiles. Limit profiles play the role of the lower and the upper bounds of the category C_h. They are ranked by the *dominance relation* $\succ_D$: $\forall i \in \{1, 2, \dots, p\}$, $b^{i+1} \succ_D b^i$ holds which means that $\forall k$, $b^{i+1}_k \geq b^i_k$ and $\exists j$, $b^{i+1}_j > b^i_j$. There are two special limit profiles, b^1 and b^{p+1} which are defined as the minimum (resp. the maximum) values: $b^1 = (b^1_1, ..., b^1_n)$, such that $\forall i$, $b^1_i = \min_{x \in X}(x_i)$ (resp. $b^{p+1} = (b^{p+1}_1, ..., b^{p+1}_n)$, such that $b^{p+1}_i = \max_{x \in X}(x_i)$).

We write $x \in C_h$ or $cl(x) = C_h$ to say that alternative x is assigned to category C_h. The set of learning alternatives is denoted by LS, while AS provides the information on the assignments of learning alternatives: $AS = \{(x, C_h), x \in LS, cl(x) = C_h\}$.

2.2 MR-Sort

MR-Sort [12, 13] is a multi-criteria ordinal classification method allowing to assign alternatives to ordered categories. The assignment procedure is based on pairwise comparisons between objects of X and limit profiles : object x is assigned to the category C_h if it is at least as good as the lower limit profile b^h and it is not at least as good as the upper limit profile b^{h+1} :

$$x \in C_h \iff x \succeq b^h \text{and } x \not\succeq b^{h+1}$$

Object x which is indifferent to the upper limit profile of the best category ($x \sim b^{p+1}$) is assigned to C_p.

The binary relation $\succeq$ is based on the weighted majority principle as in the following:

$$x \succeq b^h \iff \sum_{i:x_i \geq b_i^h} w_i \geq \lambda$$

The parameter λ is called the *majority threshold*.

3 Multiagent Architecture

We have used multiagent theory to cope with communication and collaboration issues that are crucial in any decision support system. Agents are also able to learn about their past experience and consequently to improve their decision-making process by adjusting their strategies. Thus, the agents'behaviour may change depending on the environment they are deployed on.

In addition, there are common agreed properties of agents, which include autonomy, proactiveness, responsivity and adaptability. Agents are also able to know and learn human user's preferences when they interact with them [14]. Agents typically have a partial control over their environment. Like in real life situations, each agent has a limited viewpoint and incomplete information. There is no global system control, and information is fully decentralized.

In order to address the preference elicitation problem of several users incrementally, in real time and in a distributed way, we propose a multiagent system. Thus, the first step is to specify its architecture and communication protocol by defining the structure and the actions of each agent as well as the allowed interactions between them.

Figure 1 presents our system architecture. We observe that each involved agent has a communication module which is implemented to regulate the message exchange between agents with respect to the communication protocol. Such a communication protocol specifies the types of allowed interactions and the actions that they are authorized to make during the elicitation process.

3.1 User Agent

3.1.1 Role

Each user agent communicates with a human user and learns its preferential parameters using an incremental elicitation method that will be specified in the following sections. Then,it returns the results to the mediator agent

3.1.2 Process

Each user agent *ua* is specifically dedicated to elicit a given human user's *hu* preferences by communicating with him incrementally. It has a decision module which allows to choose the "best" strategy for selecting the "best" question since we try to save elicitation effort by asking the most potentially informative question. The output of this module is the best question to ask to the human user.The latter answers the question and the response will be send to the elicitation module that will update the elicitation method (Chosen elicitation method will be discussed in the following section). We stop the procedure after k questions. At the end of the process, *ua* sends its output data to the mediator agent *ca* (Fig. 2).

3.2 Mediator Agent

3.2.1 Role

The mediator agent *ca* collects the results received from user agents, identifies similarities between them and return the results so that they can converge more quickly towards the preferences of their users.

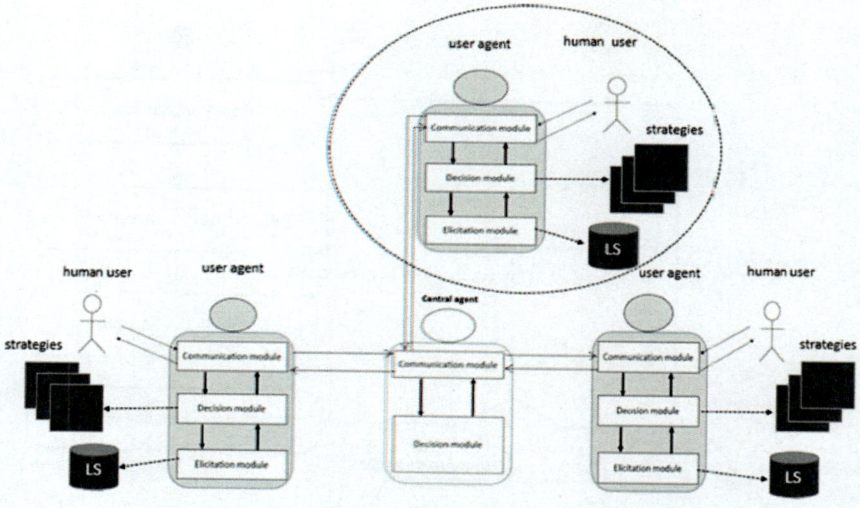

Fig. 1 Multi-agent system architecture

3.2.2 Process

During the elicitation process, each *ua* sends a request to the mediator agent to assist them in detecting similarities with other user agents. The mediator agent has a decision module that allows it to compute agents' similarities on the basis of particular metrics. Then, it returns the result to the *ua* so that it can converge more quickly towards the preferences of the human user he represents.

At the end of the elicitation process, each *ua* sends its final results to *cu* who updates its data consequently.

4 User Agent's Elicitation Module

In order to elicit human preferences, we choose as preference elicitation method the incremental MR-Sort assuming that preferences can be modeled by MR-Sort. Our elicitation procedure starts with a small learning set to which we add one by one new assignment examples. We try to choose "new" assignment examples in a way to learn as fast as possible the user's preferences, reducing, as a result, the elicitation effort and time.

Notice that MR-Sort model is defined by the following parameters: the weights $(w_1, \ldots, w_n)$, the limit profiles $(b^1, \ldots, b^{p+1})$ and the majority threshold λ. We

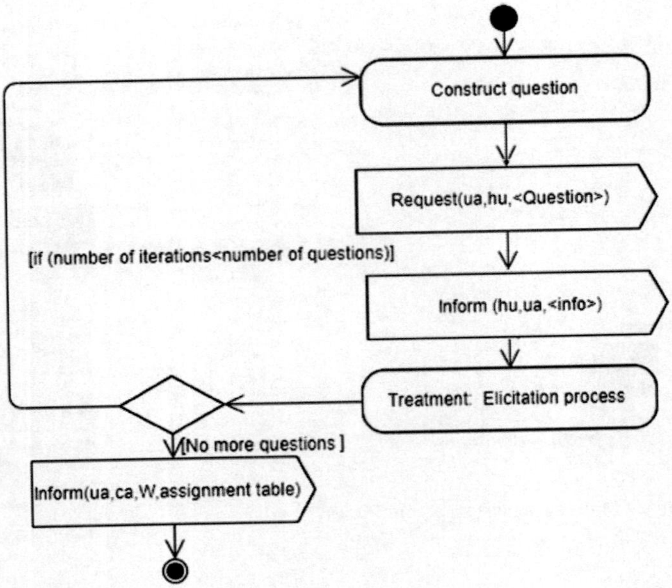

Fig. 2 User agent process

define the majority threshold and limit profiles prior to the beginning of the elicitation. The majority threshold will be fixed at 0.55 by default but may be changed if needed. Limit profiles are defined in a way that maximizes the difference $b_i^{h+1} - b_i^h, \forall i \in \{1, \ldots, n\}, \forall h \in \{1, \ldots, p\}$. Hence, the weights are the only parameters to be learned (Fig. 3).

The main steps of our approach are presented in Algorithm 1.

Algorithm 1 The MR-Sort-Inc algorithm. K is the number of questions, X the set of alternatives, C the set of categories, LS learning alternatives, AS learning assignments.

Procedure MR-Sort-Inc(K,X,C,LS,AS)
$LC \leftarrow$ linear constraints encoding AS
for $j \leftarrow 1, K$ **do**
  for $x \in X \setminus LS$ **do**
    for $C_i \in C$ **do**
      $LC' \leftarrow LC \cup \{x \in C_i\}$
      $w_{x,C_i} \leftarrow Solve\_LP(LC')$ for strategy s
    end for
    Compute $score_s(x)$
  end for
  $x^* \leftarrow \arg\max_x(score_s(x))$
  $C_{x^*} \leftarrow Answer(x^* \in ?)$
  $LC \leftarrow LC \cup \{x^* \in C_{x^*}\}$
end for
$w_{result} \leftarrow w_{x^*,C_{x^*}}$

We start with few data from the learning set (learning alternatives and their classification) and we include them as constraints in our linear program (LP). After that, we test the assignment of each object to each category using a "strategy" (different strategies will be discussed in the following section) and we define a score for each

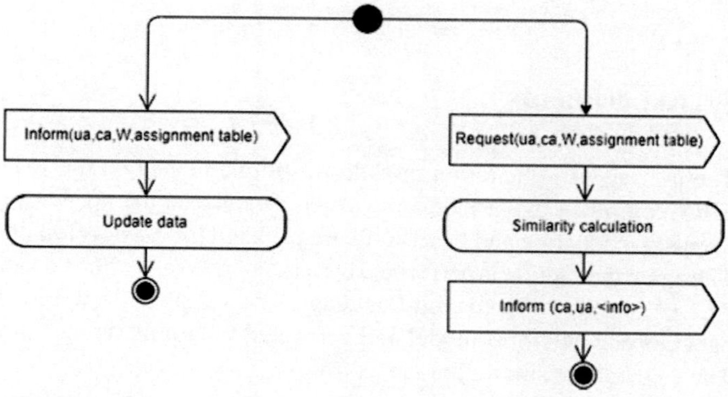

Fig. 3 Mediator agent process

object. The object having the highest score is chosen: the user is asked about the assignment of that object and the learning set is updated consequently. This procedure goes on until a stop condition is reached (in our experiments when we reach a fixed number of questions, but in real applications termination may be left to the user). The last linear program contains all the assignments of the computed learning set. Hence it provides the weights that we are looking for.

4.1 Example of Strategy to Find the "Best" Question to Ask : Number of Doubts

In the following, we present an example of how, following a strategy, we select the next question to ask to the user. This example illustrates an application of Algorithm 1. With this strategy, we aim to ask questions about alternatives that present more uncertainty about their assignment to possible categories:

- We start by data from the learning set and we add them as constraints in our linear program (LP).
- We choose an alternative x and add the assignment of x to the category C_i as a new constraint for the LP. We apply the function Solve_LP which tests the feasibility of the LP by adding this constraint. If LP is feasible we increment the number of doubts. We iterate this step by testing the assignment of the alternative x to all possible categories to have its final score representing the number of categories to which x could be assigned, we denote this score by $score_{doubt}(x)$.
- Step 2 is iterated to calculate the final $score_{doubt}$ of all alternatives in the set $X \setminus LS$. We ask the question about the assignment of the alternative maximizing the $score_{doubt}$. The user's response is then added as a fixed constraint in our initial LP.
- We stop the procedure after K questions. The weights found by the last LP are then used in order to predict the assignments of the remaining alternatives.

5 Experiments

The user agent's elicitation module was implemented on the JADE platform. It is a FIPA-compliant agent platform, which establishes the normative rules for a set of agents to cooperate and has a software package for the development of JAVA agents. The agent interaction is performed by sending messages whose content is expressed in ACL(Agent Communication Language). We conducted experiments based on input data close to MR-Sort model and generated randomly using a uniform distribution. The steps of the simulation are as follows:

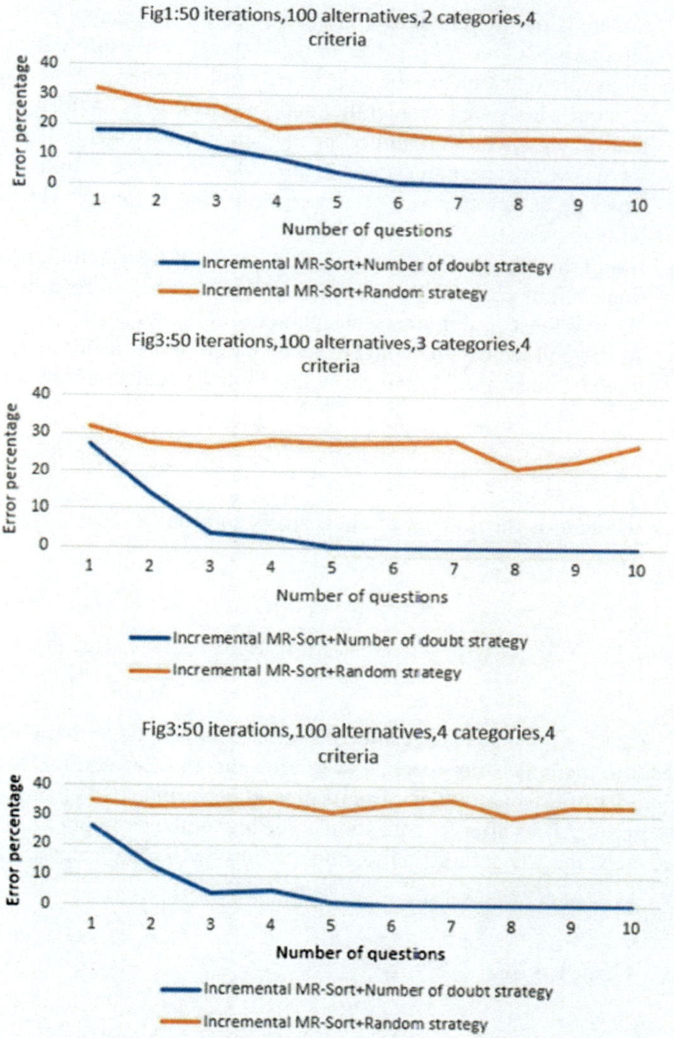

Fig. 4 $Error_1$ based on number of questions with $m = 100$,$n = 4$, $p = 2$ (right),$p = 3$ (middle),$p = 4$ (left),$|LS| = 1$ and 50 iterations

i. Using a uniform distribution for each of the parameters, we generate the performance table of alternatives, the performance table of profiles (we assume that the scale of each criterion is between 0 and 10), a vector of weight w, the majority threshold λ. We apply MR-Sort method in order to find the assignments of the generated alternatives. We call these assignments, assignments of *model 1*.

ii. We apply our incremental elicitation Algorithm 1 using the assignments of some alternatives of *model 1* as learning assignments. We apply incremental MR-Sort by asking questions to the user based on the strategy of the paper compared by random whose we consider the random selection of an object x from X (We use this strategy in order to underline the performance of Number of doubt strategy), the answer to these questions is also taken from the results of the *model 1*. We generate as output a vector of weight w' that represents the preferences of the human user.

iii. Using the limit profiles, majority threshold of our elicitation procedure and the final weight vector w', we provide the assignments of the remaining alternatives. We call these assignments, assignments of *model 2*.

iv. At the end we calculate an error rate based on the difference of assignments of *model 1* and *model 2*. We chose one metric to calculate the error rate:

$$Error_i = \sum_{x \in X} error_i(x)/m$$

where m is the number of alternatives and $i \in \{1, 2\}$.

$$error_1(x) = \begin{cases} 0 \text{ if } cl_{model_1}(x) = cl_{model_2}(x) \\ 1 \text{ else.} \end{cases}$$

Figure 4 presents the results of our simulations. Note that convergence with the random method is not reached even after the 10th question. On the other hand the Number of doubt strategy presents a good performance. We converge to the human user preferences after 5–6 questions. The higher the number of categories, the slower we converge, since there is more uncertainty with more categories.

6 Conclusion

In this article, we presented our multi agent architecture to elicit the preferences of multiple users using our Preference elicitation method: the incremental MR-Sort. We detailed the algorithm that minimizes the number of asked questions to deduce the weights of the decision model of each user. We conducted an experimental analysis to illustrate the proposed approach. In future works, we aim to implement new agent strategies and to conduct a comparative experimental analysis of several elicitation methods and test different agent behaviors (selfish agents versus altruistic agents, homogeneous population versus heterogeneous population,etc.).

References

1. Sobrie, O., Mousseau, V., Pirlot, M.: Learning the parameters of a non compensatory sorting model. In: International Conference on Algorithmic Decision Theory, pp. 153–170. Springer (2015)
2. Sobrie, O., Mousseau, V., Pirlot, M.: A population-based algorithm for learning a majority rule sorting model with coalitional veto. In: International Conference on Evolutionary Multi-Criterion Optimization, pp. 575–589. Springer (2017)
3. Teso, S., Passerini, A., Viappiani, P.: Constructive preference elicitation by setwise max-margin learning. arXiv:1604.06020, 2016
4. Teso, S., Passerini, A., Viappiani, P.: Constructive preference elicitation for multiple users with setwise max-margin. In: International Conference on Algorithmic Decision Theory, pp. 3–17. Springer (2017)
5. Guo, S., Sanner, S., Bonilla, E.V.: Gaussian process preference elicitation. In: Advances in Neural Information Processing Systems, pp. 262–270 (2010)
6. Garcia, I., Pajares, S., Sebastia, L., Onaindia, E.: Preference elicitation techniques for group recommender systems. Inf. Sci. **189**, 155–175 (2012)
7. Tyler, L., Boutilier, Craig: Robust approximation and incremental elicitation in voting protocols. IJCAI **1**, 287–293 (2011)
8. Woodridge, M., Jennings, N.R.: Intelligent agents: theory and practice the knowledge engineering review (1995)
9. Brigui-Chtioui, I., Caillou, P., Negre, E.: Intelligent digital learning: agent-based recommender system. In: Proceedings of the 9th International Conference on Machine Learning and Computing, pp. 71–76. ACM (2017)
10. Brigui-Chtioui, I., Caillou, P.: Multidimensional decision model for classifying learners: the case of massive online open courses (MOOCS). J. Decis. Syst. **26**(2), 170–188 (2017)
11. Nefla, O., Öztürk, M., Viappiani, P., Brigui-Chtioui, I.: Interactive elicitation of a majority rule sorting model with maximum margin optimization. In: International Conference on Algorithmic Decision Theory, pp. 141–157. Springer (2019)
12. Bouyssou, D., Marchant, Thierry: An axiomatic approach to noncompensatory sorting methods in MCDM, ii: more than two categories. Euro. J. Oper. Res. **178**(1), 246–276 (2007)
13. Leroy, A., Mousseau, V., Pirlot, M.: Learning the parameters of a multiple criteria sorting method. In: Algorithmic Decision Theory—Second International Conference, ADT 2011, Piscataway, NJ, USA, Oct 26–28, 2011. Proceedings, pp. 219–233 (2011)
14. Jennings, N.R., Wooldridge, M.: Applications of intelligent agents. In: Agent Technology, pp. 3–28. Springer (1998)

Automatic Grading of Knee Osteoarthritis on the Kellgren-Lawrence Scale from Radiographs Using Convolutional Neural Networks

Sudeep Kondal, Viraj Kulkarni, Ashrika Gaikwad, Amit Kharat, and Aniruddha Pant

Abstract The severity of knee osteoarthritis is graded using the 5-point Kellgren-Lawrence scale where healthy knees are assigned grade 0, and the subsequent grades 1–4 represent increasing severity of the affliction. Although several methods have been proposed in recent years to develop models that can automatically predict the Kellgren-Lawrence grade from a given radiograph, most models have been developed and evaluated on datasets not sourced from India. These models fail to perform well on the radiographs of Indian patients. In this paper, we propose a novel method using convolutional neural networks to automatically grade knee radiographs on the Kellgren-Lawrence scale. Our method works in two connected stages: in the first stage, an object detection model segments individual knees from the rest of the image; in the second stage, a regression model automatically grades each knee separately on the Kellgren-Lawrence scale. We train our model using the publicly available Osteoarthritis Initiative dataset and demonstrate that fine-tuning the model before evaluating it on a dataset from a private hospital significantly lowers the corresponding mean absolute error. Additionally, we compare classification and regression models built for the same task and demonstrate that regression outperforms classification.

Keywords Knee osteoarthritis · Kellgren-Lawrence grades · X-ray · Convolutional neural networks · Mask RCNN · Densenet-121 · Transfer learning

1 Introduction

Knee osteoarthritis is a painful joint disorder that obstructs the natural movement of the knee. Its major indications include joint space narrowing, osteophytes formation, and sclerosis [1]. It is most commonly observed in people who are above the age of 45, suffer from obesity, or lead a sedentary lifestyle. Since no known cure exists

S. Kondal (✉) · V. Kulkarni · A. Gaikwad · A. Kharat · A. Pant
DeepTek Inc., Pune, India
e-mail: sudeep.kondal@deeptek.ai

L. Troiano et al. (eds.), *Advances in Deep Learning, Artificial Intelligence and Robotics*, Lecture Notes in Networks and Systems 249,
https://doi.org/10.1007/978-3-030-85365-5_16

for reversing knee osteoarthritis, early diagnosis is crucially important for a patient to prevent its further progression by making behavioral and lifestyle changes [2]. Despite the availability of advanced medical imaging techniques such as Magnetic Resonance Imaging (MRI) and Computed Tomography (CT), the knee radiograph remains the most widely used modality for the diagnosis of osteoarthritis. The severity of the disease is graded on the Kellgren-Lawrence (KL) scale [1] where grade 0 suggests no radiographic features of OA are present; grade 1 suggests doubtful joint space narrowing (JSN) and possible osteophytic lipping; grade 2 suggests definite osteophytes and possible JSN on anteroposterior weight-bearing radiograph; grade 3 suggests multiple osteophytes, definite JSN, sclerosis, possible bony deformity; and grade 4 suggests large osteophytes, marked JSN, severe sclerosis and definite bony deformity [3].

Although it is widely used, the KL grading system suffers from reader subjectivity in evaluating radiographs [4–6]. Computer-aided diagnosis can mitigate this subjectivity by providing an automated assessment to assist radiologists in making the final decision. In this study, we propose a method for building a convolutional neural network (CNN) system to automatically grade a knee radiograph on the KL scale. We first develop a preprocessing network that takes as input a knee radiograph and outputs two images: one image containing only the left knee, and another image containing only the right knee. The second network takes as input a single knee image and outputs the predicted KL grade. For this second network, we consider two options - classification and regression - and observe that regression performs better than classification.

Most prior work on automatically diagnosing knee osteoarthritis is based on datasets which have been collected from patients not based in India, particularly the Osteoarthritis Initiative (OAI) Dataset[7–9] containing 4447 images and the Multicenter Osteoarthritis Study (MOST) Dataset [8–10] with 3026 images. We demonstrate that the model developed using OAI Dataset did not perform well on a dataset sourced from a private hospital in India (hereafter referred to as *Target Dataset*). However, when the model was fine-tuned on Target Dataset, the performance improved significantly.

In this paper, we propose a novel method using convolutional neural networks to automatically grade knee radiographs on the Kellgren-Lawrence scale. We show the use of transfer learning to adapt a model trained on one domain for a different domain with varying image characteristics. We also compare classification and regression models built for this same task and demonstrate that regression outperforms classification. The rest of the paper is organized as follows: Sect. 2 surveys prior work related to this problem; Sect. 3 describes the data and method in detail; Sect. 4 discusses the results of the experiments; and, in Sect. 5, we make concluding remarks and mention future work.

2 Related Work

Machine learning [11] has been widely used to detect and grade knee osteoarthritis. Kotti et al. [12] built a regression model on a small dataset of 94 subjects using data points such as mean value, push-off time, and slope of vertical, anterior–posterior, and medio-lateral ground reaction forces measured using force plates with piezoelectric 3-component force sensors. Bandyopadhyay et al. [13] argued that knee osteoarthritis should be diagnosed using features of the cartilage; they segmented the cartilage from the knee x-ray image, extracted features like statistical moments, texture, shape, etc., and used them to train a random forest [14] classifier to predict knee osteoarthritis severity with a classification accuracy of 87.92%. Brahim et al. [15] used images from OAI Dataset to predict knee osteoarthritis severity using image preprocessing techniques like circular Fourier filter and multivariate linear regression and extracted important features using independent component analysis; these extracted features were further used to train naive Bayes [16] and random forest classifiers. Orlov et al. proposed Wndchrm [17], an image classifier that uses hand-crafted features based on polynomial decomposition, contrast, pixel statistics, and textures in addition to features obtained by performing image transformations [17–19].

Manually designing and selecting relevant features, however, require a significant degree of domain knowledge [20], while image processing techniques such as those in [13] are not robust and are prone to noise. Convolutional neural networks (CNN) [21], on the other hand, learn relevant features automatically from the training images without human intervention, and they have achieved state-of-art performance on many image classification tasks in recent times [22]. Antony et al. [7, 8] investigated the use of well-known CNNs such as the VGG 16-layer net [23], VGG-M-128 [24], and BVLC reference CaffeNet [25] to classify knee radiographs and demonstrated that they performed better than the prevailing state-of-art methods that used hand-crafted features. Tiulpin et al. [9] extracted knee joints by generating region proposals and scoring them using a support vector machine based model; in the next step, a Siamese CNN [26] was used to predict the KL grade from these extracted patches. They also presented attention maps highlighting the radiological features affecting the network's decisions.

Machine learning methods, in general, and neural networks, in particular, are known to face difficulties in generalizing to datasets other than the ones used to train them. We show in this study that the model trained on OAI Dataset performed poorly when used to evaluate images from Target Dataset. We further show that the performance increased dramatically when the model was subsequently fine-tuned on Target Dataset. Two approaches have been considered by prior research work: the classification approach where the different KL grades 0–4 are treated as discrete, unordered classes; and the regression approach where the KL grade is treated as a continuous response variable that takes a value in the range [0, 4]. We contrast the performance of both approaches and conclude that regression performs better than classification. Our proposed method is graphically described in Fig. 1.

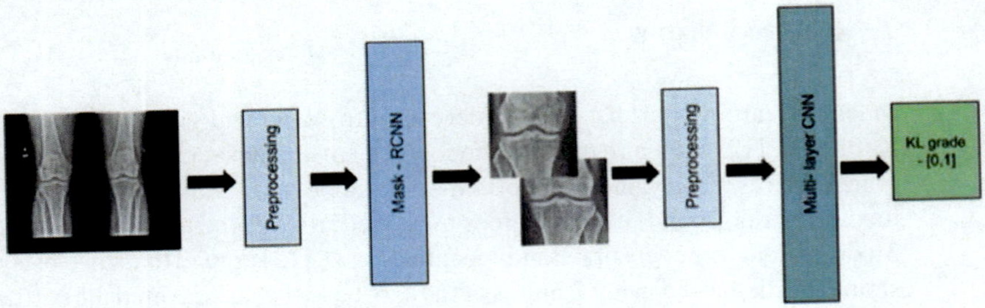

Fig. 1 Proposed pipeline for the prediction of KL grades from radiographs

Table 1 Distribution of images into KL grades

KL grade	0	1	2	3	4	Total
OAI dataset	3493	2319	1595	1177	310	8894
Target dataset	335	150	199	194	297	1175

3 Data and Method

In this section we describe the datasets, the methods used to build the models, and the results.

3.1 *Data*

We obtained a set of 4447 knee radiographs in DICOM format along with the KL grades for both left and right knees from OAI. The dataset contained images of males and females between 45 and 79 years of age. From a well-known private hospital in India, we obtained Target Dataset consisting of 1043 knee radiographs. Both datasets contained images in the anterior-posterior view. Table 1 shows the distribution of samples in different KL grades for the two datasets.

3.2 *Segmenting the Knee Joints from the Radiographs*

We randomly selected 1000 images from OAI Dataset. Using the VGG Image Annotator (VIA) tool [27], we drew bounding boxes around each knee joint in these images. These 1000 images containing 2000 annotated knee joints were used to train a Mask R-CNN model [28] that could segment the two knee joints from a given radiograph. In addition to segmenting the knees, the model was also trained to dif-

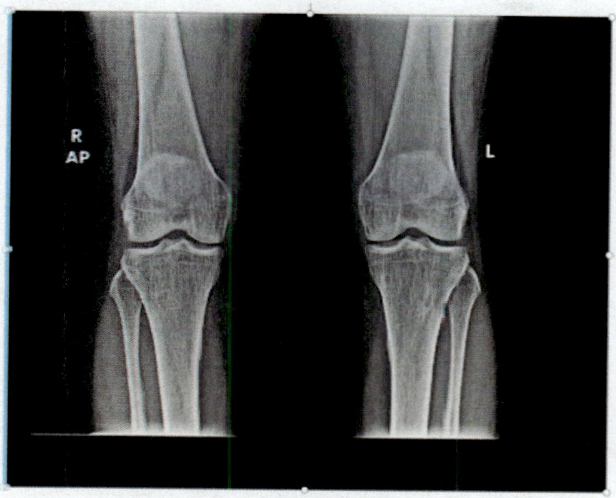

Fig. 2 Knee radiograph sample provided as input to knee segmentation model

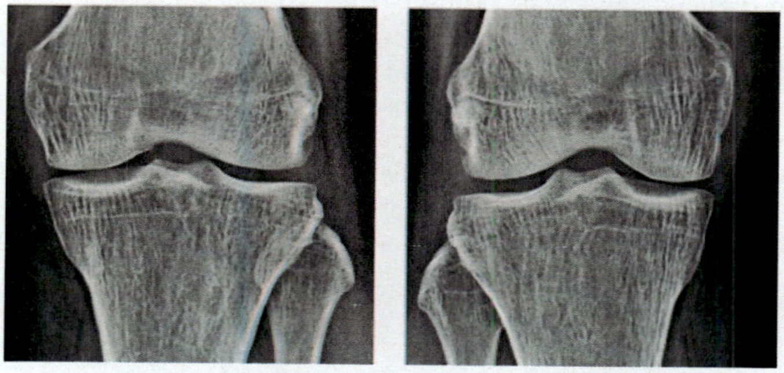

Fig. 3 Left and right knees as segmented from image in Fig. 2

ferentiate between the left and the right knee to aid the subsequent process of report generation. The Mask R-CNN algorithm works in two steps. In the first step, it generates region proposals through a region proposal network (RPN) [29]. In the second step, it predicts the class-label for each region and regresses the coordinates of the bounding box that encloses the knee joint. The model was tested on 100 randomly selected images from OAI Dataset. The mean squared error (MSE) for bounding box regression was 0.0507, the DICE score for segmenting the knees was 0.93, and the accuracy for classifying the segmented knee into left and right was 99%. Using this model, we segmented the knee joints from the rest of the images in OAI dataset (see Figs. 1, 2 and 3).

Table 2 Performance of classification model on OAI dataset when evaluated on OAI dataset

KL grade	Precision	Recall	F1-score	Kappa
0	0.67	0.88	0.76	0.57
1	0.57	0.01	0.02	0.02
2	0.63	0.47	0.54	0.41
3	0.68	0.54	0.60	0.55
4	0.18	0.95	0.31	0.27
Average	0.55	0.57	0.45	0.36

3.3 *Predicting KL Grade from Segmented Knee Images*

The segmented knee images were normalized using min-max normalization to ensure the pixel intensity values were bounded in the range [0,255]. OAI Dataset was divided into training, validation, and testing subsets in the ratios 70:10:20. Grading on the KL scale is based on joint space narrowing, and the differences between two consecutive grades are very subtle in appearance. In a conventional CNN, each layer is connected only to its adjacent layer. In DenseNet [30], however, every layer is connected to every other layer; this allows the network to learn better features from lesser data. We used the DenseNet-121 architecture to train a classification model that classified each input knee image into one class from the set of discrete, unordered classes: {0, 1, 2, 3, 4}. Cross-entropy was used as the loss function for training the model. We obtained a mean class-wise precision of 0.55 and a mean class-wise recall of 0.57 (see Table 2 for detailed results).

The KL-grades present a natural ordering amongst them. To take advantage of this ordering, we replaced the output layer of the classification network with two fully connected layers. The penultimate layer had 128 nodes with ReLU activation, while the final output layer had a single node with linear activation. Architectures for both classification and regression are presented in Fig. 4. This model was trained using mean squared error as the loss function to predict the KL grade as a real number. The model output was rounded off to the closest integer in the range [0, 4] to obtain the final predicted grade. Basically we converted the problem from classification to regression with a hope to get better results. The mean class-wise precision was 0.84 and the mean class-wise recall was 0.82 (see Table 3 for detailed results).

3.4 *Fine-Tuning Models for Target Dataset*

When evaluated on Target Dataset, the regression model trained on OAI Dataset yielded a mean class-wise precision of 0.46 and a mean class-wise recall of 0.33. Table 4 shows that the performance of the regression model trained on OAI Dataset drops when evaluated on Target Dataset. Although analyzing the reasons for this

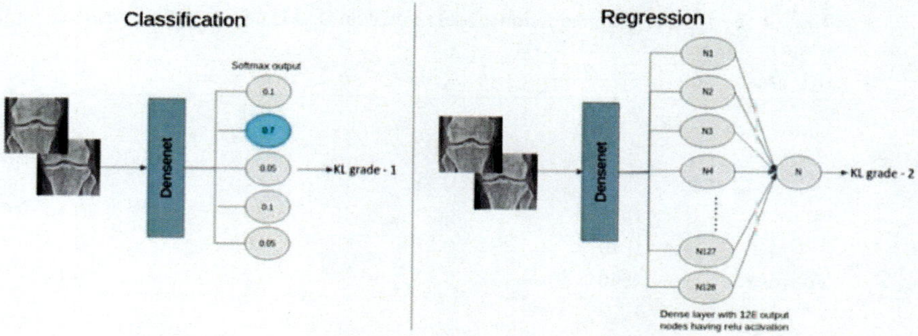

Fig. 4 Comparision between architectures used for classification and regression

Table 3 Performance of regression model on OAI dataset when evaluated on OAI dataset

KL grade	Precision	Recall	F1-score	Kappa
0	0.93	0.91	0.92	0.86
1	0.71	0.74	0.73	0.66
2	0.85	0.87	0.86	0.80
3	0.85	0.85	0.85	0.83
4	0.84	0.76	0.80	0.79
Average	0.84	0.82	0.83	0.79

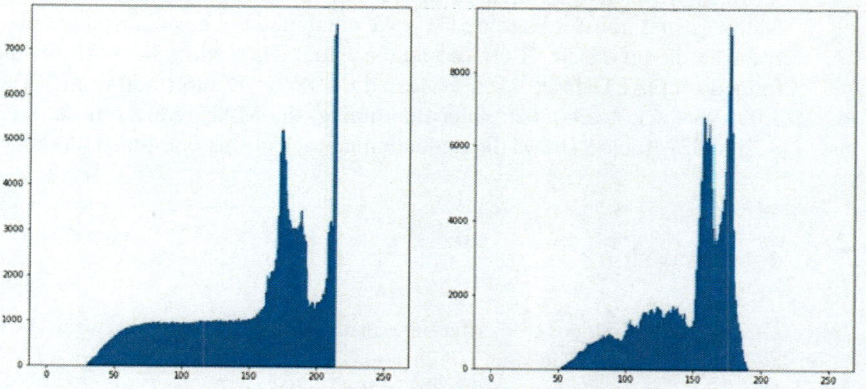

Fig. 5 (Left) Histogram of pixel intensities of images from OAI Dataset; (Right) Histogram of pixel intensities of images from Target Dataset

drop in performance is beyond the scope of the paper, we surmise that the images from the two datasets differed owing to reasons including, but not limited to, different image capture settings, different imaging equipment, and perhaps different patient population characteristics. The difference between OAI Dataset and Target Dataset is visually represented in Fig. 5.

Table 4 Performance of regression model trained on OAI Dataset and fine-tuned on Target Dataset when evaluated on Target Dataset

KL grade	Precision	Recall	F1-score	Kappa
0	1	0.02	0.04	0.03
1	0.17	0.09	0.12	0.03
2	0.22	0.43	0.29	0.09
3	0.22	0.53	0.31	0.10
4	0.70	0.57	0.63	0.52
Average	0.46	0.33	0.28	0.16

Table 5 Performance of regression model trained on OAI dataset when evaluated on target dataset

KL grade	Precision	Recall	F1-score	Kappa
0	0.90	0.79	0.84	0.79
1	0.52	0.70	0.59	0.52
2	0.63	0.68	0.65	0.58
3	0.67	0.64	0.65	0.59
4	0.91	0.86	0.88	0.85
Average	0.73	0.73	0.73	0.66

To improve the performance of the model, we fine-tuned the model on Target Dataset using transfer learning [31]. We computed the mean absolute error (MAE) between the predicted KL grade and the actual KL grade. The MAE for the model trained on OAI Dataset when evaluated on Target Dataset before fine-tuning was 1.09 (95% CI: 1.03–1.15). After fine-tuning, the MAE reduced to 0.28 (95% CI: 0.25–0.32). Table 5 shows the evaluation metrics of this fine-tuned model.

4 Discussion

Our experiments uncovered differences in image characteristics between OAI Dataset and Target Dataset. The model trained on OAI Dataset did not perform well on Target Dataset. However, when we fine-tuned the model on Target Dataset, we observed a significant improvement in performance.

Since the Kellgren-Lawrence system presupposes an implicit ordering between its grades, we argue that the grade should be treated as an ordinal variable rather than a nominal variable. Training a classification model using cross-entropy as the loss function imposes an equal penalty regardless of whether the misclassification is between neighbouring grades or far-away grades. However, defining this as a regression problem and using mean squared error as the loss function imposes lower penalties on misclassifications between neighboring grades and progressively higher penalties as the distance between the actual and predicted grades increases. When evaluating the fine-tuned model on Target Dataset, 68% of all misclassifications were amongst neighbouring grades for the classification model, while 87% of all misclassifications were amongst neighbouring grades for the regression model.

Since different stages of knee osteoarthritis require different interventions, it is important to identify the grade of osteoarthritis affecting a patient. By automatically and accurately assigning KL grades to knee radiographs, the proposed method can help (1) mitigate the effects of human subjectivity in assessing radiographs; (2) reduce radiologist work burden; (3) improve reporting times.

5 Conclusion

We demonstrate that the problem of model under-performance on a test dataset having different image characteristics than the training data can be mitigated by using transfer learning to fine-tune the model to the test dataset. We also show that treating the KL grades as a continuous response variable instead of discrete, unordered classes significantly improves the performance. In our work, we had access to labels for the test dataset. In situations where the test dataset is not labelled, unsupervised domain adaptation techniques can be considered [32–36]. We leave this for future work.

References

1. Kellgren, J., Lawrence, J.: Radiological assessment of osteo-arthrosis. Ann. Rheumatic Diseases **16**(4), 494 (1957)
2. Ji, X., Zhang, H.: Current strategies for the treatment of early stage osteoarthritis. Frontiers Mech. Eng. **5**, 57 (2019)
3. Kohn, M.D., Sassoon, A.A., Fernando, N.D.: Classifications in brief: Kellgren-Lawrence classification of osteoarthritis (2016)
4. Gossec, L., Jordan, J., Mazzuca, S., Lam, M.-A., Suarez-Almazor, M., Renner, J., Lopez-Olivo, M., Hawker, G., Dougados, M., Maillefert, J., et al.: Comparative evaluation of three semi-quantitative radiographic grading techniques for knee osteoarthritis in terms of validity and reproducibility in 1759 x-rays: report of the oarsi-omeract task force. Osteoarthritis Cartilage **16**(7), 742–748 (2008)
5. Sheehy, L., Culham, E., McLean, L., Niu, J., Lynch, J., Segal, N.A., Singh, J.A., Nevitt, M., Cooke, T.D.V.: Validity and sensitivity to change of three scales for the radiographic assessment of knee osteoarthritis using images from the multicenter osteoarthritis study (most). Osteoarthritis Cartilage **23**(9), 1491–1498 (2015)
6. Culvenor, A.G., Engen, C.N., Øiestad, B.E., Engebretsen, L., Risberg, M.A.: Defining the presence of radiographic knee osteoarthritis: a comparison between the Kellgren and Lawrence system and oarsi atlas criteria. Knee Surg. Sports Traumatol. Arthrosc. **23**(12), 3532–3539 (2015)
7. Antony, J., McGuinness, K., O'Connor, N.E., Moran, K.: Quantifying radiographic knee osteoarthritis severity using deep convolutional neural networks. In: 2016 23rd International Conference on Pattern Recognition (ICPR), pp. 1195–1200. IEEE (2016)

8. Antony, J., McGuinness, K., Moran, K., O'Connor, N.E.: Automatic detection of knee joints and quantification of knee osteoarthritis severity using convolutional neural networks. In: International Conference on Machine Learning and Data Mining in Pattern Recognition, pp. 376–390. Springer (2017)
9. Tiulpin, A., Thevenot, J., Rahtu, E., Lehenkari, P., Saarakkala, S.: Automatic knee osteoarthritis diagnosis from plain radiographs: a deep learning-based approach. Sci. Rep. **8**(1), 1–10 (2018)
10. Tiulpin, A., Thevenot, J., Rahtu, E., Saarakkala, S.: A novel method for automatic localization of joint area on knee plain radiographs. In: Scandinavian Conference on Image Analysis, pp. 290–301. Springer (2017)
11. Alzubi, J., Nayyar, A., Kumar, A.: Machine learning from theory to algorithms: an overview. J. Phys. Conf. Ser. **1142**, 012012 (2018)
12. Kotti, M., Duffell, L.D., Faisal, A.A., McGregor, A.H.: Detecting knee osteoarthritis and its discriminating parameters using random forests. Med. Eng. Phys. **43**, 19–29 (2017)
13. Bandyopadhyay, S.K., Sharma, P.: Detection of osteoarthritis using knee x-ray image analyses: a machine vision based approach (2016)
14. Breiman, L.: Random forests. Mach. Learn. **45**(1), 5–32 (2001)
15. Brahim, A., Jennane, R., Riad, R., Janvier, T., Khedher, L., Toumi, H., Lespessailles, E.: A decision support tool for early detection of knee osteoarthritis using x-ray imaging and machine learning: data from the osteoarthritis initiative. Computerized Med. Imaging Graph. **73**, 11–18 (2019)
16. Anderson, J.R., Matessa, M.: Explorations of an incremental, Bayesian algorithm for categorization. Mach. Learn. **9**(4), 275–308 (1992)
17. Orlov, N., Shamir, L., Macura, T., Johnston, J., Eckley, D.M., Goldberg, I.G.: Wnd-charm: multi-purpose image classification using compound image transforms. Pattern Recogn. Lett. **29**(11), 1684–1693 (2008)
18. Shamir, L., Orlov, N., Eckley, D.M., Macura, T., Johnston, J., Goldberg, I.G.: Source code for biology and medicine. Source Code Biol. Med. **3**, 13 (2008)
19. Shamir, L., Ling, S.M., Scott, W., Hochberg, M., Ferrucci, L., Goldberg, I.G.: Early detection of radiographic knee osteoarthritis using computer-aided analysis. Osteoarthritis Cartilage **17**(10), 1307–1312 (2009)
20. Lee, H.: Unsupervised feature learning via sparse hierarchical representations, vol. 20. Stanford University (2010)
21. LeCun, Y., Haffner, P., Bottou, L., Bengio, Y.: Object recognition with gradient-based learning. In: Shape, Contour and Grouping in Computer Vision, pp. 319–345. Springer (1999)
22. Sultana, F., Sufian, A., Dutta, P.: Advancements in image classification using convolutional neural network. In: 2018 Fourth International Conference on Research in Computational Intelligence and Communication Networks (ICRCICN), pp. 122–129. IEEE (2018)
23. Simonyan, K., Zisserman, A.: Very deep convolutional networks for large-scale image recognition. arXiv:1409.1556 (2014)
24. Chatfield, K., Simonyan, K., Vedaldi, A., Zisserman, A.: Return of the devil in the details: Delving deep into convolutional nets. arXiv:1405.3531, 2014
25. Jia, Y., Shelhamer, E., Donahue, J., Karayev, S., Long, J., Girshick, R., Guadarrama, S., Darrell, T.: Caffe: convolutional architecture for fast feature embedding. In: Proceedings of the 22nd ACM International Conference on Multimedia, pp. 675–678, 2014
26. Chopra, S., Hadsell, R., LeCun, Y.: Learning a similarity metric discriminatively, with application to face verification. In: 2005 IEEE Computer Society Conference on Computer Vision and Pattern Recognition (CVPR'05), vol. 1, pp. 539–546. IEEE (2005)
27. Dutta, A., Zisserman, A.: The VGG image annotator (via). arXiv:1904.10699, 2019
28. He, K., Gkioxari, G., Dollár, P., Girshick, R.: Mask r-cnn. In: Proceedings of the IEEE International Conference on Computer Vision, pp. 2961–2969, 2017
29. Ren, S., He, K., Girshick, R., Sun, J.: Faster r-cnn: towards real-time object detection with region proposal networks. In: Advances in Neural Information Processing Systems, pp. 91–99, 2015

30. Huang, G., Liu, Z., Van Der Maaten, L., Weinberger, K.Q.: Densely connected convolutional networks. In: Proceedings of the IEEE Conference on Computer Vision and Pattern Recognition, pp. 4700–4708, 2017
31. Tan, C., Sun, F., Kong, T., Zhang, W., Yang, C., Liu, C.: A survey on deep transfer learning. In: International Conference on Artificial Neural Networks, pp. 270–279. Springer, 2018
32. Ben-David, S., Blitzer, J., Crammer, K., Kulesza, A., Pereira, F., Vaughan, J.W.: A theory of learning from different domains. Mach. Learn. **79**(1–2), 151–175 (2010)
33. Wang, M., Deng, W.: Deep visual domain adaptation: a survey. Neurocomputing **312**, 135–153 (2018)
34. Ganin, Y., Ustinova, E., Ajakan, H., Germain, P., Larochelle, H., Laviolette, F., Marchand, M., Lempitsky, V.: Domain-adversarial training of neural networks. J. Mach. Learn. Res. **17**(1), 2030–2096 (2016)
35. Long, M., Zhu, H., Wang, J., Jordan, M.I.: Unsupervised domain adaptation with residual transfer networks. arXiv:1602.04433, 2016
36. Tzeng, E., Hoffman, J., Saenko, K., Darrell, T.: Adversarial discriminative domain adaptation. In: Proceedings of the IEEE Conference on Computer Vision and Pattern Recognition, pp. 7167–7176, 2017

Robust Model for Rural Education Using Deep Learning and Robotics

Sailesh Iyer

Abstract Rural Education is important for the overall development of villages. National Achievement Survey (NAS) has surveyed and reported in many States of India consistent decline in the learning levels of students in mathematics, language and science from class III to class VIII studying in the government school system. Smart Villages is only possible if the literacy level and infrastructure improves considerably. This paper aims to perform a reality check of the situation by comparing different rural areas of various countries including India and study of related work done. The paper proposes a Robust Model for Rural Education by developing an intelligent humanoid robot using the Deep Learning approach, Human recognition, Object Recognition and Speech Recognition. The data set consists of Primary and Secondary Student data of around 10,000 Students (5 years) from 5 villages. The Proposed Model would be compared with existing models on the parameters of Learnability, Decision making, Flexibility and Cost-effectiveness. The implementation of this Model will help in decreasing the drop out rate, evaluate Students and give them a Learning platform based on their characteristics, increase adaptive and self paced learning. This Model can also be executed for Rural Adult Education and Skill building so that the Smart Village concept can become a reality

Keywords Rural education · Humanoid robot · Deep learning · Human recognition · Object recognition · Speech recognition · Learnability · Robust model · Adaptive learning · Smart village

1 Introduction

The entire Education Scenario has seen rapid improvement raising many challenges like Technology Adaptation, Expectation of Students, Intense Competition in Industry leading to sustained employability Rural Education is even more challenging with Technology Infrastructure issues, Lack of appropriate funding, low literacy level etc. The employability of Technical Graduates is less than 30% which

S. Iyer (✉)
Rai University, Saroda Village, Ahmedabad, Gujarat 382260, India

L. Troiano et al. (eds.), *Advances in Deep Learning, Artificial Intelligence and Robotics*, Lecture Notes in Networks and Systems 249,
https://doi.org/10.1007/978-3-030-85365-5_17

is alarming. Skill Development is at a new low leading to Educational Institutions emphasising on Skill Development and Project based learning. A survey of the Rural Schools conducted as per report of National Achievement Survey (NAS) of Indian Express suggests serious gaps in the study level of the Students with less than 30% of Students could identify numeric digits above 100 [1] (Fig. 1).

Several other problems Rural Students face is distance, acute poverty, parents mindset towards education, Lack of Internet and Electricity. One more important finding was that the Rural Schools are understaffed and existing staff are used for Clerical and Administrative work. The Central Government Budget 2019 has targeted 150,000 Smart Villages ensuring Wi-fi and other Infrastructure connectivity by 2025 [2] (Fig. 2).

There have been some refreshing experiments conducted by Indus International School [3] and Keeko Robot in Beijing [4]. The implementation of Humanoid Robots

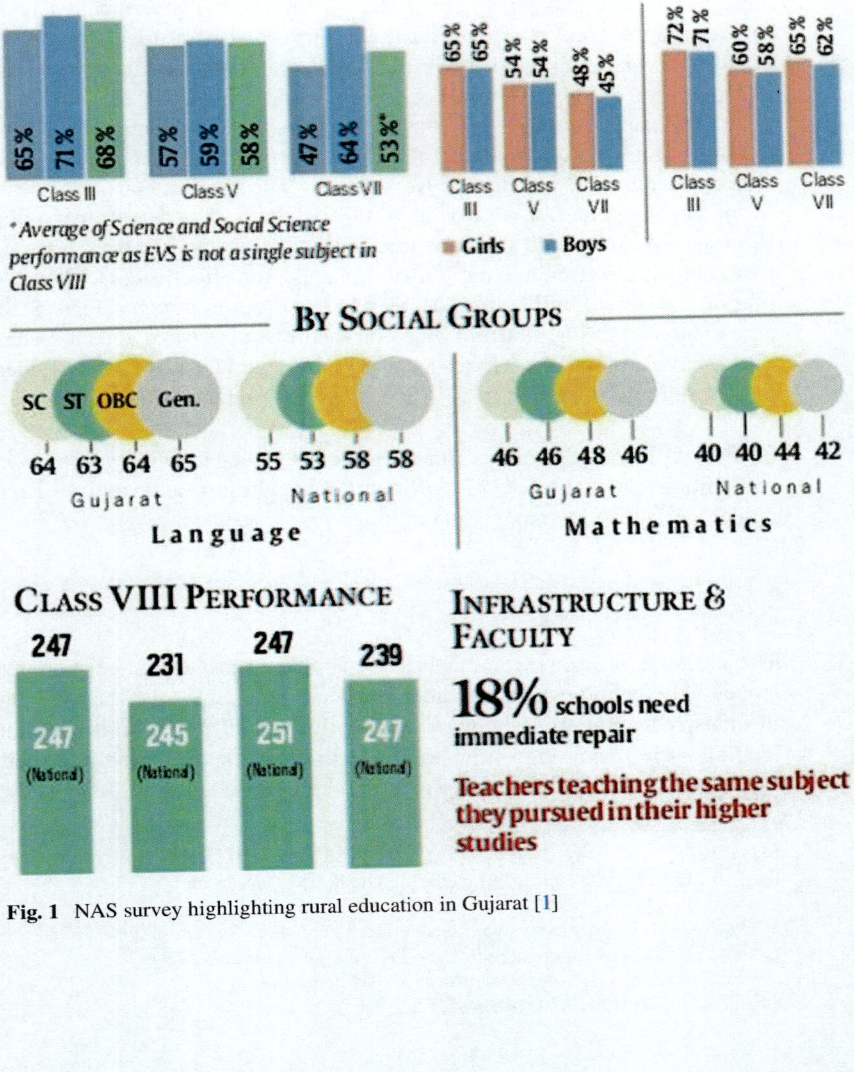

Fig. 1 NAS survey highlighting rural education in Gujarat [1]

Fig. 2 Humanoid robots assisting teacher in schools [3, 4]

in Schools [3, 4] have been encouraging with Students welcoming the concept of Humanoid Robot assisted Education and generated curiousity of the Students.

2 Related Work

Erich et al. have surveyed the different Robots used in Education and found NAO Robot to be more effective than the others [5]. Baxter et al. through experiments and survey for age group around 3 years found that Customised Robots are preferred to normal Robots [6]. Henkemans et al. concluded that using NAO Robot for age group of 7–12 years was full of knowledge and evolved learning process into one filled with fun [7]. Yi et al. involved College Students and found that the robots were effective in learning process and hands-on training [8]. Fernández-Llamas et al. observed the experience of Students in the age group of six to sixteen years and found that they preferred Robots for Teaching compared to regular Teachers [9].

Rosi et al. reported contrasting results in their experiment, targeted towards age group of 8–19 years, which found that Robots had no impact in the Teaching Learning process [10]. Huskens et al. conducted designing experiment in the age group of 8–12 years and concluded that Robot interaction induced creativity among Students [11]. Shiomi et al. found that the Robot does not increase curiosity among Students in the age group of 9–13 years [12]. Boboc discovered a new area that Robots can be used as Tour guide in Educational Institutes to attract Students [13]. Wainer et al. focussed on Teachers survey of Schools and identified that most of the Teachers are positive towards Robots assisting in Education and believe presence of Robot generates positive vibes [14]. Chin et al. used Robots with Primary Students and found that all Students wanted Robots in their class and were relaxed with Robots compared with Teachers [15]. Bacivarov and Ilian have concluded after a study of Primary Students that Humanoid robot resulted in 61% Higher participation, 80%Higher performance, and creative thinking, 0.81 Higher social impact, 100% Course attendance, and zero drops out rate [16].

Bahishti and Adam have studied the impact of using Humanoid Robot on Society and found that Technology will govern the way primary and secondary education would phase out [17].

The literature review of the work done and Traditional Education where Teachers overburdened with the Administrative, Clerical and Teaching load find it difficult to concentrate on primary objective of teaching students effectively. Inherent problems identified in Traditional Rural Education:

- Lack of Quality Education and Standard Study Materials.
- Lack of Expertise Related to Domain Area.
- Infrastructure and Connectivity Issues.
- Lack of Industry Oriented Skills and ICT Advancements.
- Lack of Basic Amenities.
- Funding Utilization.
- Understaffing and Overutilization in other Administrative Work.
- Mindset of Stakeholders.
- Low Literacy Level

Various Studies conducted have found that introducing Humanoid Robots for supporting Teachers can play a vital role in increasing Student interest and motivate students to have hands on experience as far as practical knowledge is concerned.

3 Proposed System

Extensive Review of existing models and drawbacks of Traditional Education Model formed the basis for suggesting a Robust Humanoid Robot Model which is removing the Traditional issues as well as providing cost-effective, localized and effective learning Humanoid Robot to assist in Rural Education. Proposed Model also looks at various dimensions of learning from psychological perspective of Student and keep regional diversities in mind while using technology for implementing the System.

Steps of Proposed Model are listed below:

1. Identifying Student Characteristics and Behavioural parameters. Rural Data of Students according to their age and Learning behaviour are recorded. We are concentrating on previous 5 years data to build our Model.
2. Study the Humanoid Robot market and features and select best suited to our needs.
3. Program the Humanoid Robot to identify the Learning habits based on Student face image capture and map it to the Student characteristics.
4. The Robot based on processing would suggest a learning roadmap including Adaptive Learning, Self paced learning, devising Online test according to learning pattern and assist the Teacher in providing Individual mentoring to the Student.

5. Role of Teacher would be instrumental as Teacher would analyze the results which is called Knowledge Analytics. This would be helpful in providing future roadmap for mentoring the Student.
6. Assessment Tools are devised to provide Performance Status and help in knowledge orientation and making the assessment process student centric and friendly.

The dataset consists of 5 villages near Ahmedabad, Gujarat and last 5 year records of around 10,000 Student have been collected from the Schools for the Students of 1–8th Standard. Data pre-processing has been performed and now the learner Robust Model as per steps mentioned in Fig. 3 are under progress. Humanoid Robot acquisition process is at the finalization stage. The primary focus of this model is to provide a cost-effective and efficient solution which can deal with flaws of existing models and provide a regional language induced mechanism empowering and encouraging the Student to practical learning suited to the ability and needs of the Student.

4 Results and Discussion

Various models have been proposed and implemented for Urban Schools where Robot is used just to assist the Teacher in conducting quiz or correcting assignments. This model focusses on Robot using Deep Learning Algorithm to provide a Learning roadmap and monitor the progress of the Student. The reports would be provided to the Teacher on a periodical frequency of 15 days so that the Teacher can record the progress of the Student over previous fortnight. Salient features of this Model over existing models are that it is Cost effective, encourages Student behaviour oriented learning, flexible and self paced learning.

The following are the benefits accruing from the Proposed Model:

1. Increasing concept of Adaptive and Self paced learning.
2. Interactive Tutorials using Games, Quiz and other techniques to increase the involvement of Student.
3. Identification and recognition of Learning Path of Student based on their characteristics.
4. Visually managed Interface for understanding challenging subjects.
5. Faculty Monitoring through Dashboard for monitoring fortnightly progress and evaluation of Student performance.
6. Regional language as mode of communication in addition to primary teaching language.
7. Implementing MOOC's and other open source Learning Resources in regional language to allow expert level learning by Students irrespective of their location.
8. Low cost model which can provide a big boost to Smart Village concept.

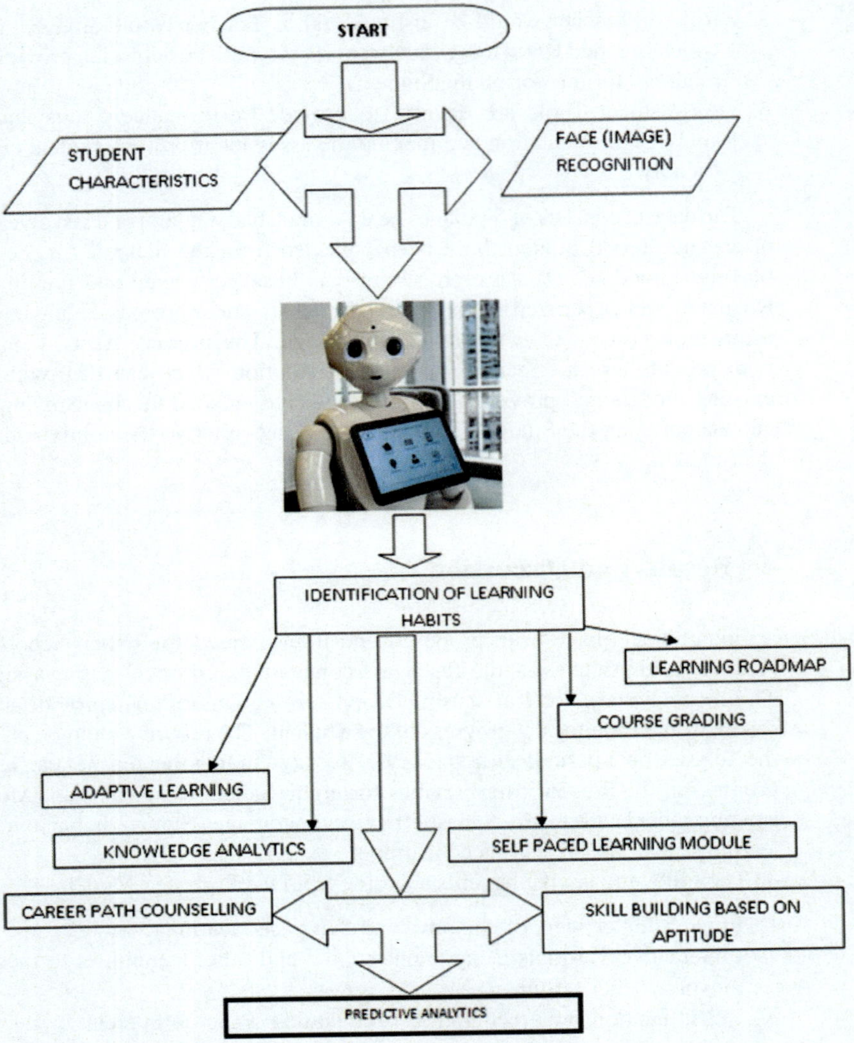

Fig. 3 Proposed robust humanoid robot model for rural education

5 Conclusions and Further Work

This results of the Comparative study of existing model revels that there is huge gap in expected parameters of Education in rural areas. Robots have been developed as an Assistant to Teaching community and fail to address the major concerns and challenges of Rural Education discussed in Sect. 4. The proposed model data collection and pre-processing of around 10,000+ has already been collected spread over past 5 years and deep learning algorithm/model is currently being developed. This proposed model prototype to be implemented by December 2021 and final adaptable version by December 2022. The proposed model provices a low cost, regional language, progress roadmap of Student monitored through Dashboard, Periodic Student Evaluation. Future work involves expanding the same model for Adult Rural Education and to map this model to various other Rural Villages making Villages self sustainable in Smart Education through low priced effective solutions which can be borne by respective Regional Governments.

Acknowledgements The author wishes to acknowledge that this is a live Rural Education project in progress and completion expected by December 2022.

References

1. https://indianexpress.com/article/education/hardlook-state-of-education-in-gujarat-part-ii-5166122/
2. https://www.deccanherald.com/business/budget-2019/budget-2019-big-plans-for-rural-india-745227.html
3. https://www.livemint.com/technology/tech-news/robots-turn-teachers-in-bengaluru-school-thanks-to-ai-1567336339210.html
4. https://www.thehindu.com/sci-tech/technology/robot-teachers-in-chinas-kindergartens/article24812421.ece
5. Erich, F., Hirokawa, M., Suzuki, K.: A systematic literature review of experiments in socially assistive robotics using humanoid robots. Cornell University, pp 1–6 (2017)
6. Baxter, P., Ashurst, E., Read, R., Kennedy, J., Belpaeme, T.: Robot education peers in a situated primary school study: personalization promotes child learning. Plos One **12**, e0178126 (2017)
7. Henkemans, O.A.B., Bierman, B.P.B., Janssen, J., Looije, R., Neerincx, M.A.: Design and evaluation of a personal robot playing a self management education game with children with diabetes type 1. Int. J. Human Comput. Stud. **106**, 63–76 (2017)
8. Yi, H., Knabe, C., Pesek, T., Hong, D.W.: Experiential learning in the development of a DARwIn-HP humanoid educational robot. J. Intell. Rob. Syst **81**, 41–49 (2016)
9. Fernández-Llamas, C., Conde, M.A., Rodríguez-Lera, F.J., Rodríguez-Sedano, F.J., García, F.: May I teach you? Students' behavior when lectured by robotic versus human teachers. Comput. Human Behav. **80**, 460–469 (2018)
10. Rosi, A., Dall'Asta, M., Brighenti, F., Del Rio, D., Volta, E.: The use of new technologies for nutritional education in primary schools: a pilot study. Public Health **140**, 50–55 (2016)
11. Huskens, B., Palmen, A., Van der Werff, M., Lourens, T., Barakova, E.: Improving collaborative play between children with autism spectrum disorders and their siblings: the effectiveness of a robot-mediated intervention based on Lego® therapy. J. Autism Dev. Disord. **45**, 3746–3755 (2015)
12. Shiomi, M., Kanda, T., Howley, I., Hayashi, K., Hagita, N.: Can a social robot stimulate science curiosity in classrooms? Int. J. Soc. Rob. **7**, 641–652 (2015)
13. Boboc, R.G.: An educational humanoid laboratory tour guide robot. Proc. Soc. Behav. Sci. 424–430 (2014)
14. Wainer, J., Dautenhahn, K., Robins, B., Amirabdollahian, F.: A pilot study with a novel setup for collaborative play of the humanoid robot KASPAR with children with autism. Int. J. Soc. Rob. **6**, 45–65 (2014)
15. Chin, K., Wu, C., Hong, Z.W.: A humanoid robot as a teaching assistant for primary education. In: Fifth International Conference on Genetic and Evolutionary Computing, pp. 21–24 (2011)

16. Bacivarov, I.C., Ilian, V.L.: The paradigm of utilizing robots in the teaching process: a comparative study. Int. J. Technol. Design Edu. **22**, 531–540 (2012)
17. Bahishti, Adam, A.: Humanoid robots and human society. Adv. J. Soc. Sci. **1**(1), 60–63 (2017). https://doi.org/10.21467/ajss.1.1

Action Recognition for Anomaly Detection Using Transfer Learning and Deep Architectures

Francisco Luque Sánchez, Siham Tabik, and Francisco Herrera

Abstract Automatic video surveillance is an emerging research area, where a huge number of publications are appearing everyday. However, there is still much room for improvements, since results of state-of-the-art are still limited. The mainstream approach to the problem using deep models consists in transfer learning from action recognition and weakly supervised fine-tuning. Nonetheless, there does not exist a systematic study of the importance of the different steps in the process. In this work, our aim is to identify the key aspects of those models and assess the importance of the different pieces that compose the final solution.

Keywords Action recognition · Anomaly detection · Deep learning · Transfer learning

1 Introduction

Automatic video surveillance is an emerging research area. Growing crime and terrorism rates and the augment of global population have produced an increasing necessity of efficient and accurate models for this task [12]. Particularly, crowd anomaly detection has proven to be a useful approach for unconstrained scenarios. This paradigm aims to learn the pattern of normal behaviors and establish a measure of similarity

F. L. Sánchez (✉) · S. Tabik · F. Herrera
Department of Computer Sciences and Artificial Intelligence, Universidad de Granada, Granada, Spain
e-mail: fluque@decsai.ugr.es

S. Tabik
e-mail: siham@decsai.ugr.es

F. Herrera
e-mail: herrera@decsai.ugr.es

L. Troiano et al. (eds.), *Advances in Deep Learning, Artificial Intelligence and Robotics*, Lecture Notes in Networks and Systems 249,
https://doi.org/10.1007/978-3-030-85365-5_18

between patterns. If a new pattern deviates much from the learned normal behavior, that pattern is marked as anomalous. However, due to the huge variability that can be present in normal videos, it is difficult to represent all the possible normal patterns in a dataset for the model training, and these approaches often suffer from high false alarm rates.

To mitigate this problem, a possible approach is to train these models in a supervised way, and thus learn both from normal and abnormal patterns. However, the problem of this approach is the need to temporally annotate the training videos. This process is costly and cumbersome, since it requires to iterate over all the frames in the dataset and indicate the label (normal or anomalous) for each one. Moreover, the publicly available databases are usually small, which is a serious problem when training deep models [7].

In order to overcome this limitations, a recent trend is the combination of transfer learning from action recognition with a weakly supervised fine-tuning for anomaly detection. Using transfer learning, the size of the anomaly detection datasets can be greatly reduced, and the public datasets for action recognition are big enough to pretrain deep learning models [1, 11].

The weakly supervised fine tuning refers to the employment of video-level annotations instead of annotations at frame level. Using this approach, the amount of expert time required for labeling the dataset is greatly reduced, since only one label per video is required. However, a specific training policy is needed, since our aim is to learn to recognize anomalies at frame level using only information at video level.

Taking into account this discussions, the purpose of this study is to assess the importance of the different decisions taken when designing an anomaly detection system, in order to identify possible shortcomings in current models. We propose a standard pipeline for analysing deep models using transfer learning and weakly supervision. An experimental study of the importance of the different decisions in the pipeline is conducted.

The rest of the work is organized as follows. In Sect. 2, related topics to this contribution are briefly discussed. In Sect. 3 the proposed experimental framework is explained. In Sect. 4 the current results are shown. Finally, in Sect. 5 conclusions derived from the study are explained and possible future research directions are proposed.

2 Related Works

In this section, the main topics in which our work relies are described. A brief explanation of those topics and an analysis of the state of the art are performed.

2.1 Deep Learning Based Action Recognition

Action recognition is a subtask of video classification, in which our system aims to assign a label to each video in our dataset, using a set of predefined classes that represent human behaviors. It is a particular case of a more general task called video classification, where the predefined classes are general and not limited to videos where people are shown. This problem has been widely studied, and a lot of works have tried to solve the task using classical models [8].

The irruption of Deep Learning in the field produced groundbreaking improvements in the results. Due to this fact, there has been a huge trend to propose deep learning models for action recognition in the last few years [4].

A video can be seen as a sequence of images with a strong temporal relation between subsequent frames. Then, leveraging temporal information in the predictions usually improves the results. To this end, the combination of CNNs for spatial features extraction and RNNs (specially LSTM models) for temporal aggregation of features was one of the first attempted approaches, and many models have been proposed [16]. A similar approach is the use of 3D CNNs, which are an extension of classical CNNs that also apply convolution in the temporal dimension. To the best of our knowledge, the first proposal of 3D CNNs is found in [15].

Another typical trend consists in combining independent models for spatial and temporal predictions. These models often employ CNNs over raw frames for spatial information extraction, and CNNs over stacked maps of optical flow for temporal information. Plenty of examples of this models can be found in the literature [1, 10, 17].

The main problems of the listed models are their high computational requirements. Most of those models are computationally expensive, and thus they are not suitable for real world applications. Last efforts focus on maintaining the classification capabilities of previous models while reducing significantly the amount of computation needed. A large number of works propose modifications in the models listed previously to this end [2, 6].

2.2 Video Anomaly Detection

Crowd anomaly detection is a task that aims to identify anomalous events performed by humans from video footage. In contrast with action recognition, the main objective in this task will be to distinguish between normal, expected behaviors and situations not expected a priori.

This problem has attracted a lot of attention in the last few years, and the number of publications addressing the topic increases every day. An extensive review about this problem can be found in [7].

2.3 Weakly Supervised Learning

The problem of video anomaly detection has been tackled using very different approaches. One of the first attempts to solve the problem consists of training the models using a supervised approach [13]. The training dataset is composed of normal and abnormal videos, where every frame is labeled as normal or abnormal, or where each video is only composed either by normal or abnormal frames. The main problem with this approach is the amount of time required for labeling, since every frame have to be properly marked. Due to this fact, there are not big datasets properly labeled in this way.

Another trend consists of training the models using a one class unsupervised paradigm. In this case, the training set is formed only by normal footage, and the system learns to characterize the patterns in normal videos. Then, comparing those patterns with the features extracted in the test set, an anomaly score is produced. Examples of this approach are [3], where a Gaussian model over the deep extracted features is learned, or [9], where the reconstruction error of video snippets is used as an anomaly score. The problem with this approach is produced by the variability present in normal scenes. Since the different settings that can be found in a normal video are huge, it is difficult to properly represent all that information on a training dataset. Due to this fact, a normal pattern that deviates much from the learned patterns over the training set can easily occur, and thus a false alarm is produced.

To solve this problems, weakly supervised learning is proposed. Weakly supervision in video assumes that only video level annotations are available for training. However, the untrimmed nature of the videos produce that anomalous sequences contain also normal footage. Due to this fact, special strategies for training the models must be developed, which can be found in the literature.

Sultani et al. [14] proposed a multiple instance learning approach. In this approach, videos are split into segments, and each set of segments compose a bag of samples with only one label per bag. For normal bags, it is clear that all the samples will belong to the normal class. For abnormal bags, some of the samples will belong to the abnormal class, while other ones will be normal. The optimization process then aims to maximize the difference in the anomalous score output by the model between the most anomalous segment in abnormal bags and the most anomalous segment in normal bags. By maximizing this margin, the network learns to predict low anomaly scores for normal segments and high scores for abnormal ones.

Another approach can be found in [18]. In this case, the problem is formulated as a problem of learning under noisy labels. Two models are involved in the training procedure. The action recognition model, modified for anomaly classification, predicts the anomaly score for video segments. On the other side, a graph convolutional network [5] corrects the prediction of low confidence segments using temporal and appearance similarity of high confidence segments. Both models are trained until convergence.

3 Pipeline for Weakly Supervised Crowd Anomaly Detection

In this section, the experimental framework employed in our experimentation is proposed.

The key piece of the framework is the pipeline explained in this section. Using this pipeline, works that perform crowd anomaly detection using weak supervision can be summarized by specifying three different aspects. Afterwards, the dataset employed for evaluation of the anomaly detection task is explained. Finally, the different choices taken in our experiments are commented.

3.1 Pipeline

After a thorough review of the works that solve the problem of video anomaly detection using weakly supervision, we identified a common pipeline shared among most of the works:

1. An action recognition model is selected.
2. The model is pretrained using an action recognition dataset.
3. The pretrained model is used for feature extraction.
4. An anomaly detector model is fine-tuned over the extracted features using weak supervision.
5. The final model is evaluated over the test subset.

Given the previous standardization, all the solutions can be described specifying three elements:

- The action recognition model.
- The action recognition dataset for pretraining.
- The weakly supervision policy for fine-tuning.

The schema of the proposed pipeline is outlined in Fig. 1. Given this simplification of the solutions, our next aim is to assess the importance of the decisions taken in the different points of the pipeline, that is:

- How does the selected model affect the final performance?
- Does the final results depend on the pretraining dataset?
- Is the fine-tuning policy impactful?

3.2 Anomaly Detection Dataset

The anomaly detection dataset used as benchmark is the UCF-Crime Dataset [14]. This dataset is a collection of untrimmed videos taken from surveillance video cam-

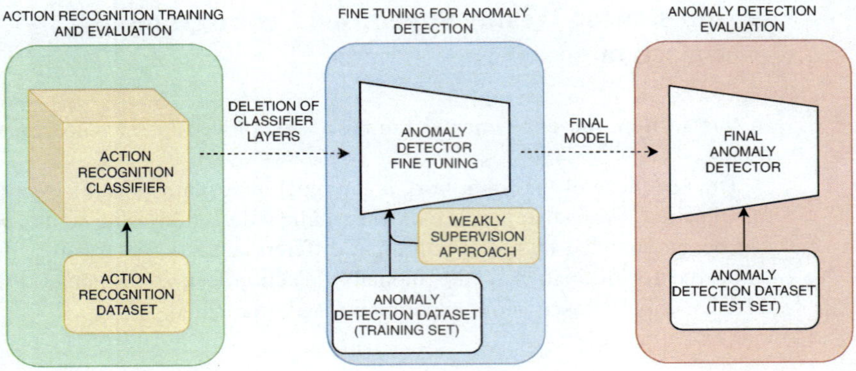

Fig. 1 Pipeline proposed. Elements marked in yellow are the possible stages affecting performance. Best viewed in color

Fig. 2 Examples of different frames in UCF-Crime Dataset. A huge variability of contexts, light conditions and camera angles are present. Moreover, the source of anomaly in videos is also diverse, so our models should be able to generalize properly and adapt to different environments and situations. Best viewed in color

eras in different settings. In Fig. 2 several frames of the dataset are shown as examples. As it can be observed, the diversity of environments that can be found in the dataset is important. Also, sources of anomaly are diverse, which forces our system to perform correctly under very different situations.

The dataset is composed of 1900 videos, evenly distributed in normal and abnormal. The training set consists of 800 normal and 810 abnormal videos, while the test set includes the remaining 150 normal and 140 anomalous videos. Anomalies are further classified into 13 different classes, but they will be ignored in this work, since we are interested in the anomaly detection, not in the further classification.

3.3 *Choices in the Pipeline*

As it has been stated before, we are interested in measuring the impact of each decision over the pipeline. To this end, we have performed a series of experiments varying the chosen approach in each of the stages. The different elements we have worked with are the following:
Action recognition dataset We have pretrained our models in two different datasets for action recognition:

- UCF101 Dataset [11]: Contains 13,320 videos from 101 classes.
- Kinetics-400 Dataset [1]: Contains 306,245 videos from 400 classes.

Action recognition classifier The action recognition classifiers used as feature extractors are the following:

- C3D [15]: This model uses a 3D convolutional neural network to extract features from video snippets.
- Temporal Segment Networks (TSN) [17]: This model employs a combination of convolutions in subsequent frames, using RGB and optical flow maps as inputs, to extract information from different parts of the video snippet.
- I3D [1]: This approach creates a 3D convolutional network by inflating 2D kernels, leveraging pretraining in image datasets.
- Temporal Shift Module [6]: Features extracted from previous frames are shifted to subsequent frames during feature extraction, to enrich feature maps with previous information.
- SlowOnly [2]: The video is processed by several networks, using a frame rate dependent of the depth of each architecture. In this model, only the deepest network is employed.

Fine-tuning policy Two different policies for fine tuning have been explored:

- Multiple Instance Learning [14]: This fine tuning policy proposes to divide videos in snippets, and maximize the distance between the maximum anomaly score in a normal video and an anomalous one. To this end, an specific loss function based on the Hinge loss together with smoothness and sparsity restrictions is proposed.
- Graph Convolutional Noise Cleaning [18]: This method proposes an alternative optimization framework. It alternates the training of the anomaly detector and a graph convolutional net, which learns to group video snippets using temporal adjacency and feature similarity.

4 Experimental Results

In this section, an experimental study using the listed choices is performed, in order to validate the proposed pipeline. For every combination of *pretraining dataset—action recognition model—fine tuning policy*, the following metrics are calculated; Top-1 and Top-5 accuracies in the pretraining dataset, which gives an idea of the model capabilities; the area under the ROC curve and the average precision in the anomaly detection dataset, which is indicative of the final model performance detecting abnormal situations; and the false alarm rate in the normal videos in the test set of the anomaly detection dataset, which indicates the robustness of the model ignoring normal situations in different environments.

The obtained results are summarized in Table 1.

As we can observe, this framework greatly simplifies the comparison between approaches. For example, taking into account the results we observe in the table, we can extract several conclusions about the performance of the tested models:

– Models that employ TSN as action recognition model perform robustly, independently of the pretraining dataset and fine-tuning policy. This behavior is shown by the stability of the metrics in the anomaly detection task.

Table 1 Obtained results depending on the decisions in the pipeline

Fine-tuning policy	Dataset	Model	Top-1 pret.	Top-5 pret.	AUC	Avg. precision	FAR
MIL	UCF-101	C3D	83.29	95.64	70.20	14.08	3.23
		I3D	81.52	96.05	76.29	17.96	0.18
		SlowOnly	83.32	94.77	75.26	16.74	0.53
		TSN	82.18	97.09	77.98	18.00	0.12
		TSM	85.04	96.67	75.47	17.64	0.30
	Kinetics400	C3D	61.13	85.24	75.19	20.02	27.00
		I3D	72.85	90.75	77.03	17.83	1.41
		SlowOnly	75.55	92.04	79.18	20.99	0.49
		TSN	71.80	90.17	77.93	18.74	3.17
		TSM	70.25	89.66	75.78	18.74	3.31
GCN	UCF-101	C3D	83.29	95.64	71.91	15.28	4.16
		I3D	81.52	96.05	75.69	16.71	1.79
		SlowOnly	83.32	94.77	74.34	16.69	0.95
		TSN	82.18	97.09	78.14	17.91	2.30
		TSM	85.04	96.67	75.92	17.18	0.77
	Kinetics400	C3D	61.13	85.24	75.08	19.41	25.59
		I3D	72.85	90.75	78.13	19.46	2.53
		SlowOnly	75.55	92.04	78.06	22.65	2.89
		TSN	71.80	90.17	77.17	17.33	4.60
		TSM	70.25	89.66	75.51	20.55	3.99

All results are expressed in percentage. For every metric except FAR, higher is better. In case of FAR, lower is better

- C3D models are strongly affected by pretraining dataset, suggesting that this model is less robust when learning features, despite being one of the most employed models in the literature.
- Both fine-tuning policies tested yield similar results for every model. This situation can be indicative of the less impact of this step in the pipeline.
- A correlation between performance at pretraining and performance at anomaly detection can be found. Specially, top-5 accuracy at pretraining is a good proxy for final performance.
- Dataset is impactful. Most models benefit from training on a bigger dataset, since results when using Kinetics 400 for pretraining are consistently better than the obtained when pretraining with UCF-101.
- There is still a huge room for improvement over the positive class. Results can be interpreted as fairly good in terms of the AUC, which is the usual metric employed in this problem, but observing the average precision of the system at positive class results are far from satisfactory.
- The false alarm rate calculated over normal videos is fairly low, which indicates that our models are capable of properly ignoring normal situations, and generate alarms only for abnormal videos. This situation indicates robustness of the models against changes in environment.

5 Conclusions

In this work, an experimental framework for crowd anomaly detection using transfer learning and weak supervision is outlined. Using the proposed pipeline, models that employ transfer learning and weakly supervision for fine tuning can be described in terms of three different parameters; the pretraining dataset, the action recognition model, and the weakly supervision approach. Extensive experiments using the pipeline are performed, obtaining several conclusions over the impact of each stage.

Derived from the current study, several questions appear as possible paths for future work, which aim at validating the conclusions extracted in this work; can the models benefit from even bigger datasets? For the moment, only two datasets have been tested, but there are bigger datasets available, which can lead to better results; can more intelligent fine-tuning policies improve the results further? The policies tested in our case lead to similar results for most of the models, but different results using other pretraining policies are expected.

Acknowledgements This work was supported by the Project AI-MARS (CIEN CDTI Programme, IDI-20181108). Francisco Luque Sánchez and Siham Tabik were supported by the Spanish Government fellowship programmes Formación de Profesorado Universitario (FPU19/04640) and Ramón y Cajal (RYC-2015-18136) respectively.

References

1. Carreira, J., Zisserman, A.: Quo vadis, action recognition? A new model and the kinetics dataset. In: Proceedings of the IEEE Conference on Computer Vision and Pattern Recognition, pp. 6299–6308 (2017)
2. Feichtenhofer, C., et al.: Slowfast networks for video recognition. In: Proceedings of the IEEE International Conference on Computer Vision, pp. 6202–6211 (2019)
3. Feng, Y., Yuan, Y., Lu, X.: Learning deep event models for crowd anomaly detection. Neurocomputing **219**, 548–556 (2017)
4. Herath, S., Harandi, M., Porikli, F.: Going deeper into action recognition: a survey. Image Vis. Comput. **60**, 4–21 (2017)
5. Kipf, T.N., Welling, M.: Semi-supervised classification with graph convolutional networks. arXiv preprint arXiv:1609.02907 (2016)
6. Lin, J., Gan, C., Han, S.: TSM: temporal shift module for efficient video understanding. In: Proceedings of the IEEE International Conference on Computer Vision, pp. 7083–7093 (2019)
7. Luque-Sánchez, F., et al.: Revisiting crowd behaviour analysis through deep learning: taxonomy, anomaly detection, crowd emotions, datasets, opportunities and prospects. Inf. Fusion (2020)
8. Poppe, R.: A survey on vision-based human action recognition. Image Vis. Comput. **28**(6), 976–990 (2010)
9. Ramchandran, A., Sangaiah, A.K.: Unsupervised deep learning system for local anomaly event detection in crowded scenes. Multimed. Tools Appl. 1–21 (2019)
10. Simonyan, K., Zisserman, A.: Two-stream convolutional networks for action recognition in videos. Adv. Neural Inf. Process. Syst. 568–576 (2014)
11. Soomro, K., Zamir, A.R., Shah, M.: UCF101: a dataset of 101 human actions classes from videos in the wild. arXiv preprint arXiv:1212.0402 (2012)
12. Sreenu, G., Saleem Durai, M.A.: Intelligent video surveillance: a review through deep learning techniques for crowd analysis. J. Big Data **6**(1), 48 (2019)
13. Sudhakaran, S., Lanz, O.: Learning to detect violent videos using convolutional long short-term memory. In: 2017 14th IEEE International Conference on Advanced Video and Signal Based Surveillance (AVSS), pp. 1–6. IEEE (2017)
14. Sultani, W., Chen, C., Shah, M.: Real-world anomaly detection in surveillance videos. In: Proceedings of the IEEE Conference on Computer Vision and Pattern Recognition, pp. 6479–6488 (2018)
15. Tran, D., et al.: Learning spatiotemporal features with 3D convolutional networks. In: Proceedings of the IEEE International Conference on Computer Vision, pp. 4489–4497 (2015)
16. Ullah, A., et al.: Action recognition in video sequences using deep bidirectional LSTM with CNN features. IEEE Access **6**, 1155–1166 (2017)
17. Wang, L, et al.: Temporal segment networks: towards good practices for deep action recognition. In: European Conference on Computer Vision, pp. 20–36. Springer (2016)
18. Zhong, J.-X., et al.: Graph convolutional label noise cleaner: train a plug-and-play action classifier for anomaly detection. In: Proceedings of the IEEE Conference on Computer Vision and Pattern Recognition, pp. 1237–1246 (2019)

Personalised Federated Learning with BERT Fine Tuning. Case Study on Twitter Sentiment Analysis

Jose Antonio Ruiz-Millán, Eugenio Martínez-Cámara, M. Victoria Luzón, and Francisco Herrera

Abstract Federated learning is a machine learning approach where many clients jointly train a model under the orchestration of a central server, while keeping the data sequestered on the clients. It allows to preserve data privacy and to reach a higher learning generalisation capacity. Also, the global model of federated learning has to adapt to the particularities of the data of the clients. We propose the use of fine tuning as a personalisation method for adapting a global federated learning model to the data of new clients. We use sentiment Twitter data as case of study, and we show that the use of fine tuning improve the performance of the global model and of the local model of the clients.

Keywords Personalised federated learning · Fine tuning · Sentiment analysis

1 Introduction

Artificial Intelligence (AI) methods need data to accomplish their goals, standing out deep learning methods that are highly data eager methods. Currently, users demand personalised AI services, which may require the use of their personal data. However,

J. A. Ruiz-Millán (✉) · E. Martínez-Cámara · M. Victoria Luzón · F. Herrera
Andalusian Research Institute in Data Science and Computational Intelligence, University of Granada, Granada, Spain
e-mail: jantonioruiz@ugr.es

E. Martínez-Cámara
e-mail: emcamara@decsai.ugr.es

M. Victoria Luzón
e-mail: luzon@ugr.es

F. Herrera
e-mail: herrera@decsai.ugr.es

L. Troiano et al. (eds.), *Advances in Deep Learning, Artificial Intelligence and Robotics*, Lecture Notes in Networks and Systems 249,
https://doi.org/10.1007/978-3-030-85365-5_19

they are reluctant to share their personal data to central servers, due to their unceasing increasing awareness about protecting their privacy. Hence, we need a learning paradigm that is able to learn from data of different users and to preserve data privacy.

Federated learning (FL) is a nascent machine learning paradigm where many clients, in the sense of electronic devices or entire organisations, jointly train a model under the orchestration of a central server, while keeping the training data decentralised in their data silos [13]. The joint model or the FL model is iteratively built only using the model parameters of the learning models of the clients. Thus, FL is able to develop AI services that preserve data privacy, because the data are kept sequestered by the data owners or clients.

Since FL is able to use more data to build learning models, these ones will have a higher generalisation capacity. However, the global FL model has to take into account the particularities of the data distribution of the clients, and even more when there are clients that were not participated in the training of the global FL model. Hence, the FL model has also to be able to be personalised to meet the particularities of the clients data [13].

The personalisation of the FL model resembles to adapt a learning model to a specific domain, an for this reason transfer learning is used as a method for personalising FL models [1]. Multi-task learning also attains to transfer knowledge from a set of tasks to a target one, and it is thus applied in FL to transfer knowledge from some clients to new ones [17]. On the other hand, the particularities of FL may recommend to develop specific personalisation methods for FL, but the need of having some grade of knowledge from the clients data may push to not strictly follow the FL definition, through sharing a short amount of data between the clients an the server as in [18].

The use of language is a particular feature of each person. Speakers of the same language usually show a particular use of the language, for instance Spanish speakers from Spain and America subtly differ in the use of the Spanish language. Hence, Natural Language Processing tasks (NLP), as Sentiment Analysis (SA) in Twitter, are tasks that need some grade of personalisation in FL.

In this paper, we focus on the data of new clients incorporated to the FL setting. We propose to follow a fine tuning approach to transfer the knowledge from the FL model, and to adapt it to the particularities of the data of those new clients. We build our fine tuning personalised FL setting upon the BERT pre-trained language model [8]. We first train the model on subsets of clients (training clients), and we evaluate the FL model on a different set of clients (testing clients), which only use a reduced set of data to fine tune the FL model.

We use the Sentiment Analysis in Twitter as a case study task, and we evaluate our fine tuning personalised FL setting on the FL setting of the Sentiment140 dataset of the LEAF FL benchmark [3]. We compare our fine tuning personalisation approach with other two FL setting with any personalisation layer. The results show that personalisation based on fine tuning allows to enhance the performance of new incorporated clients, and to reduce the performance divergences amongst them.

The rest of the paper is organised as follows: we succinctly review in Sect. 2 the most recent works related to personalised federated learning. Section 3 is devoted to fine tuning for personalising federated learning, and we define the experimental setup in Sect. 4. Then, we show the results of the experimentation in Sect. 5, and we highlight the conclusions of this paper in Sect. 6.

2 Background on Personalised Federated Learning

PFL is the specific FL setting for adapting the FL global model to the particularities of clients data in inference time [13]. There are some scenarios that require the personalisation of the FL model, namely: (1) when the data distribution of the clients are non-IID; (2) when the feature space of the clients data are different; (3) when there are differences in the label space on the data of the clients; (4) when there are differences in the feature and label space and (5) when there is a need to incorporate new clients in different times of the FL model life.

We find two approaches for addressing personalisation in FL: (1) the adjustment of methods used for domain adaptation or transferring knowledge from one task to other one, and (2) the design of new specific methods for personalising the FL model.

2.1 Existing Machine Learning Approaches Adapted to FL

The machine learning approaches used for domain adaptation or transferring knowledge are leverage to provide personalisation capacity to FL. We subsequently highlight the approaches used so far in the literature:

Multi-Task Learning [4]: It allows to use the knowledge from one task into a target task allowing to enhance the performance on the target task. This approach is leverage in FL to transfer the knowledge from a set of clients, which are considered as tasks, into the clients in which the FL model has to be personalised. In [6], the authors propose *VIRTUAL*, a method for federated multi-task learning with non-convex models, where the federated setting is set as a star-shaped Bayesian network, and the learning is performed using approximated variational inference. The *MOCHA* method is proposed in [24], which applies the multi-task learning approach to solve a FL problem.

Meta Learning [20]: It is similar to multi-task learning, since it also learns how to perform a tasks by using the experience from other tasks. In FL, it considers each client as an independent tasks, so it uses the learning of each client to personalised the FL model. *FedMeta* algorithm is proposed in [5] to replace FedAvg [19] for adapting MetaSGD [15] or MAML [11] to FL providing some grade of personalisation. In [10], the authors propose *Per-FedAvg*, an algorithm that apply the meta learning formulation to solve a FL problem.

Knowledge Distillation [12]: It proposes to distil the knowledge from one learning model to another, so it is adapted to the FL setting by means the distillation of the

FL model to the clients model that need to be personalised. The *FedMD* model is proposed in [14], and it uses an unique model for each client, it thus uses a dataset shared amongst the server and all the clients. Moreover, it uses the predictions of each client to aggregate the models of the clients on the server. The *FedDF* method is proposed in [17], it is similar to *FedMD*, however, it uses the model parameters to aggregate the FL model, and then it uses the predictions on the shared dataset to retrain all the clients. We remark that the previous approaches do not strictly follow the FL definition, because they share the data amongst the server and the clients.

Transfer Learning [21]: It is used in domain adaptation problems for transferring the knowledge from one task to other target task. Thus, it is used in FL for transferring the knowledge from a set of clients to other clients. The *FedPer* method is proposed in [1], and it provides a personalisation layer for deep feed forward neural networks in FL. The personalisation layer is not shared with the FL server, while the rest of the architecture of the deep feed forward neural network does. The *LG-FedAvg* model [16] follows an inverse approach of *FedPer*, because the personalisation layer is shared with the FL server, and the other layer of the deep learning model are kept in the clients.

2.2 *Personalisation Methods Specifically Designed for FL*

We find in the literature some methods specifically design for providing personalisation capacity to FL. Some of those methods are:

Hypercluster [18]: It clusters the clients of the FL setting and then keep a FL model per each cluster of clients, hence this method provides a personalisation capacity per each cluster of clients.
pFedme [9]: The personalisation is performed in the clients, by means adjusting with a regulation term the update of the client learning models with the FL model.
APFL [7]: It follows a similar approach than in *pFedme*, but in this case each client dynamically sets the level of update of the learning models of the clients.
DAPPER [18]: It proposes a personalisation method that does not strictly follow the definition of FL, because it consists in training the learning models of the clients on a dataset composed of the union of the shared global data and the local data of each client.

3 Fine Tuning for Personalising Federated Learning

Fine tuning allows to transfer the underline knowledge of a dataset to a learning model trained on other dataset. Accordingly, fine tuning is usually used to leverage pre-trained models in large datasets to resolve machine learning tasks that lack of

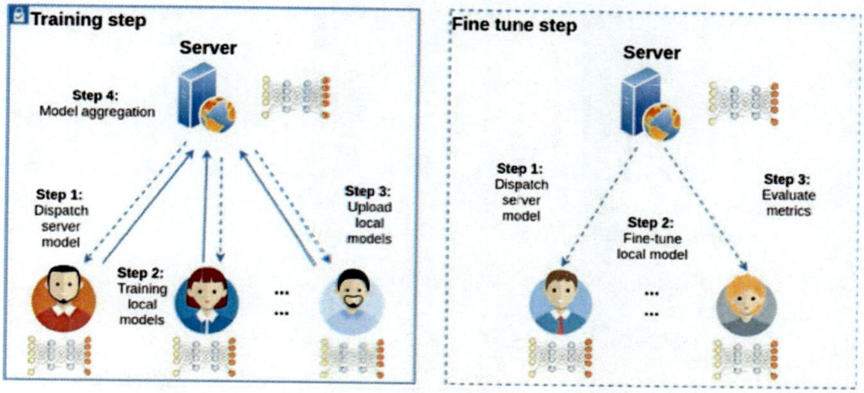

Fig. 1 The PFL setting. On the left, the training of the FL model. On the right the fine tune step of new clients

large datasets. Accordingly, it can be used to personalise FL models to the data of the clients, and to integrate clients with few training data in the FL setting. We propose to use fine tuning approach in the former scenario.

The PFL based on fine tuning requires of a pre-trained global model (PGM), a subset of clients to train (training clients) a specific downstream task, which in our case is sentiment analysis (SA) in Twitter, and a subset of clients to test the trained model. Accordingly, the PFL setting will consist on: (1) broadcasting the PGM to the training clients and running as much rounds of learning [23] as needed; and (2) deploying the trained PGM on the testing clients, which only have to perform a single training round on a small subset of their data to personalised the PGM to the particularities of their data. Figure 1 depicts the PFL setting based on fine tuning.

4 Experimental Setup

We perform two experiments in order to evaluate our PFL based on fine tuning. In the first experiment, we set 10 clients as training clients, and we set 6 clients as training clients in the second experiment. In both experiments, we set 6 clients as new clients or test clients, since we assess (1) if fine tuning is able to adapt the trained FL model to the new clients using very few data from those new clients, and (2) if the personalisation depends on the number of training clients.[1]

We use the Sentiment140 dataset from the FL benchmark LEAF [3] for the evaluation. It is a balanced dataset of 1,600,000 positive and negative tweets for binary polarity classification. We distribute about the 2.5% of the dataset (36,000 tweets) into the test clients, and the rest of the tweets were used by the training clients. The

[1] The number of clients in the evaluation was defined because of computational power restrictions.

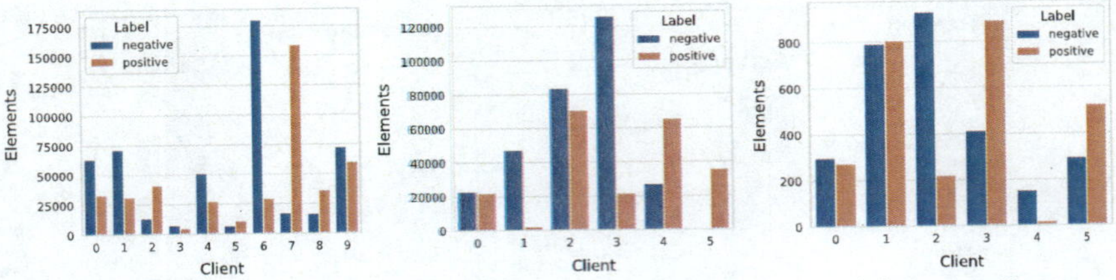

(a) 10 training clients. (b) 6 training clients. (c) 6 test clients.

Fig. 2 Data distribution for training and test clients

Table 1 Parameters used in the experiments, where lr is the learning rate

Model	Batch size	Epochs	Rounds	Optimizer
Feed-forward	64	5	20	Adam()
BiLSTM	264	3	10	Adam()
BERT-based PFL	32	1	5	Adam(lr = 2e−5)

data on the training clients is split into 80% for training and the 20% for assessing the performance of the training clients. On the other hand, the test clients only used 20% of their data to fine-tune the FL model, and the 80% of the data for testing.

We distributed the data amongst the training and testing clients following a non-iid distribution, which is shown in Fig. 2. We stress out that the data distribution shows one of the challenges of PFL in particular and FL in general, which refers to the possibility that some of the clients may lack examples of any of the classification labels. For instance, the client 5 of Fig. 2b does not have any negative tweet.

We compare our PFL based on fine tuning with two FL baseline models, namely:

Feed-forward neural network. It is built upon a word embeddings look up layer, a fully-connected layer activated with the ReLu function and a sigmoid classification layer. It uses a set of pre-trained word embeddings trained on the Google News 200B corpus [2].

BiLSTM neural network. As the previous model, it uses a word embeddings look up layer, which is followed by a dropout layer, a CNN layer, a Bidirectional Long Short-Term Memory layer, two full-connected layers activated with the ReLu function with a Dropout layer between them and a sigmoid clasification layer. It uses the set of pre-train word embeddings GloVe [22].

We use the BERT language model as the pre-train learning model for our fine tuning based PFL, and we call it BERT-based PFL model. Hence, it consists in two layers: (1) the BERT language models and (2) a sigmoid classification layer that fine tunes BERT to the SA classification task. Table 1 shows the parameters used in the experiments for each model.

We use accuracy as the evaluation measure, since it is broadly used in SA and other text classification tasks. We assess the performance of the PFL model in three learning moments: (1) when the client models are aggregated in the server using an independent test subset from the Sentiment140 dataset (acc. g1); (2) when the clients are updated with the aggregated model of the server using the test subset of the clients (acc. g2); and (3) after a training round of fine tuning in the clients (acc. l.).

5 Experimental Results

We present in this section the results of our evaluation. We run the three FL models 6 times,[2] and we show the average of the evaluation measures.

We first show the results reached by the models in the training phase, so the test clients are not used. Figure 3 shows the results of ten training clients scenario. In this case, the non-BERT models have a similar performance, while our BERT-based PFL model reached the highest results in the three evaluation measures since the first round of training. Figure 4 shows the results when the number of training clients is six. In this case, there is an additional difficulty, which is the existence of clients that lack of tweets of one of the classification labels. This fact is behind the fluctuations in the results, specially in the performance of the BiLSTM neural network model. The BERT-based PFL model again outperforms the two baselines in the three evaluation measures, and its behaviour has less fluctuations.

Figure 5 shows the results reached by each of the test clients, i.e the clients that need the personalisation of the FL model, when the number of training clients is 10. The results show a similar trend than in the previous evaluation, an our BERT-based PFL model outperforms the two baselines. Moreover, the BERT-based PFL model allows that the six text clients have a similar behaviour. Figure 6 shows the results of each of the six text clients when the number of training clients is six. The BERT-based PFL model again outperforms the two baselines, and the six text clients have a more stable performance in comparison with the other two baselines.

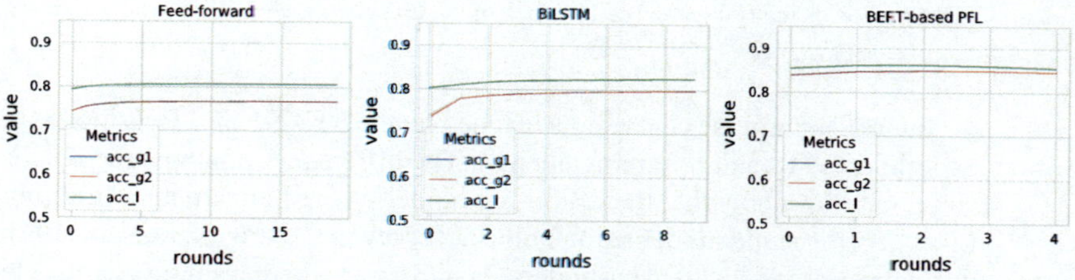

(a) Feed-forward. (b) BiLSTM. (c) BERT-based PFL.

Fig. 3 It shows the results reached by each FL model in the training step of the experiment with ten clients

[2] The number of executions of each experiment is constrained to computational restrictions.

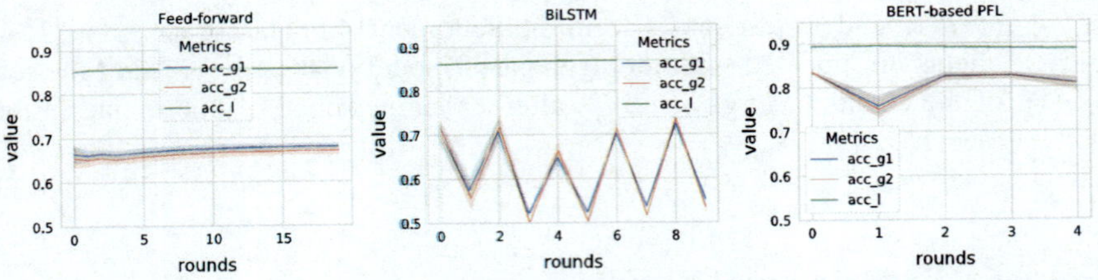

(a) Feed-forward. (b) BiLSTM. (c) BERT-based PFL.

Fig. 4 It shows the results reached by each FL model in the training step of the experiment with six clients

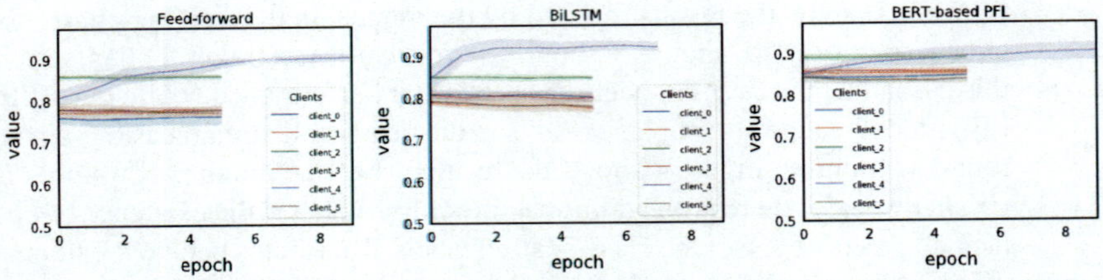

(a) Feed-forward. (b) BiLSTM. (c) BERT-based PFL.

Fig. 5 It shows the results reached by each of the text clients when there are ten training clients

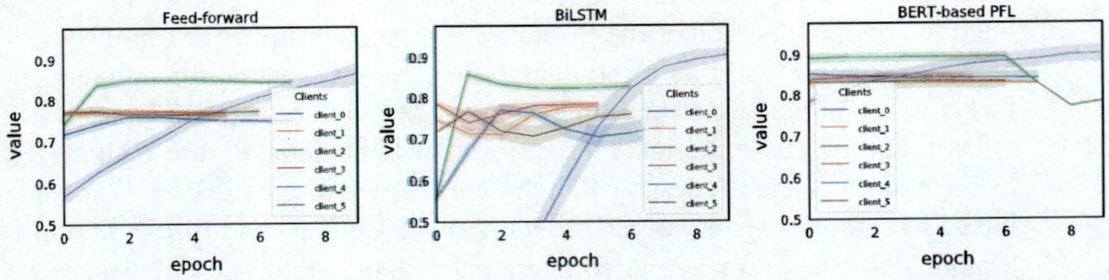

(a) Feed-forward. (b) BiLSTM. (c) BERT-based PFL.

Fig. 6 It shows the results reached by each of the text clients when there are six training clients

Table 2 summarises the results reached by the BERT-based PFL model in the two experiments (10 and 6 training clients). The differences amongst the results are not large, but they show that the results are higher when there are more training clients. Likewise, the standard deviation of the accuracy in the test clients (acc. l.) shows the differences amongst them is shorter when there are more training clients. Hence, the number of training clients influences in the performance of the personalisation.

The results show that the use of fine tuning as a personalisation method for new clients in the FL setting enhances the performance of those new clients. Likewise, we have shown that the number of training clients influences the performance of the personalisation.

Table 2 Results reached evaluating new clients

Model	acc. g1	acc. g2	acc. l.
BERT FT-10	0.834 (±0.00208)	0.844 (±0.00200)	0.867 (±0.00216)
BERT FT-6	0.831 (±0.00164)	0.838 (±0.00108)	0.858 (±0.00315)

6 Conclusions

We claim that the generalisation capacity of FL models needs to be personalised or adapted to the data of the clients, which is crucial when new clients are incorporated to the FL setting. Accordingly, we propose a PFL setting based on fine tuning. We assess the PFL setting on two different training cases and a test case, and we compare it with two deep learning models on the Twitter Sentiment Analysis task. The results show:

1. The use of pre-trained language models in FL improve the performance of the FL model.
2. When there are clients with different labels, the performance of all models worsens.
3. The number of training clients and data is relevant to the results.
4. The performance on the clients increases when a training round is performed (fine tune) before the prediction.

Therefore, we conclude that fine tuning can be used to personalised FL models.

Acknowledgements This research work is partially supported by the contract OTRI-4137 with SHERPA Europe S.L., the Spanish Government project TIN2017-89517-P. Eugenio Martínez Cámara was supported by the Spanish Government fellowship programmes Juan de la Cierva Incorporación (IJC2018-036092-I). The experiments described in this paper were developed with the Sherpa.ai FL framework. https://github.com/sherpaai/Sherpa.ai-Federated-Learning-Framework.

References

1. Arivazhagan, M.G., Aggarwal, V., Singh, A.K., Choudhary, S.: Federated learning with personalization layers. CoRR abs/1912.00818 (2019)
2. Bengio, Y., Ducharme, R., Vincent, P., Janvin, C.: A neural probabilistic language model. J. Mach. Learn. Res. **3**, 1137–1155 (2003)
3. Caldas, S., Duddu, S.M.K., Wu, P., Li, T., Konečný, J., McMahan, H.B., Smith, V., Talwalkar, A.: Leaf: a benchmark for federated settings. CoRR abs/1812.01097 (2018)
4. Caruana, R.: Multitask learning. Mach. Learn. **28**(1), 41–75 (1997)
5. Chen, F., Luo, M., Dong, Z., Li, Z., He, X.: Federated meta-learning with fast convergence and efficient communication. CoRR abs/1802.07876 (2019)
6. Corinzia, L., Buhmann, J.M.: Variational federated multi-task learning. CoRR abs/1906.06268 (2019)
7. Deng, Y., Kamani, M.M., Mahdavi, M.: Adaptive personalized federated learning. CoRR abs/2003.13461 (2020)

8. Devlin, J., Chang, M.W., Lee, K., Toutanova, K.: BERT: pre-training of deep bidirectional transformers for language understanding. In: Proceedings of the 2019 Conference of the North American Chapter of the Association for Computational Linguistics: Human Language Technologies, pp. 4171–4186 (2019)
9. Dinh, C.T., Tran, N.H., Nguyen, T.D.: Personalized federated learning with Moreau envelopes. CoRR abs/2006.08848 (2020)
10. Fallah, A., Mokhtari, A., Ozdaglar, A.: Personalized federated learning: a meta-learning approach. CoRR abs/2002.07948 (2020)
11. Finn, C., Abbeel, P., Levine, S.: Model-agnostic meta-learning for fast adaptation of deep networks. In: Proceedings of the 34th International Conference on Machine Learning, Proceedings of Machine Learning Research, vol. 70, pp. 1126–1135 (2017)
12. Hinton, G., Vinyals, O., Dean, J.: Distilling the knowledge in a neural network. In: NIPS Deep Learning and Representation Learning Workshop (2015)
13. Kairouz, P., McMahan, H.B.: Advances and open problems in federated learning. Found. Trends® Mach. Learn. **14**(1) (2021)
14. Li, D., Wang, J.: FedMD: heterogenous federated learning via model distillation. CoRR abs/1910.03581 (2019)
15. Li, Z., Zhou, F., Chen, F., Li, H.: Meta-SGD: learning to learn quickly for few-shot learning. CoRR abs/1707.09835 (2017)
16. Liang, P.P., Liu, T., Ziyin, L., Allen, N.B., Auerbach, R.P., Brent, D., Salakhutdinov, R., Morency, L.P.: Think locally, act globally: federated learning with local and global representations. CoRR abs/2001.01523 (2020)
17. Lin, T., Kong, L., Stich, S.U., Jaggi, M.: Ensemble distillation for robust model fusion in federated learning. CoRR abs/2006.07242 (2020)
18. Mansour, Y., Mohri, M., Ro, J., Suresh, A.T.: Three approaches for personalization with applications to federated learning. CoRR abs/2002.10619 (2020)
19. McMahan, H.B., Moore, E., Ramage, D., Hampson, S., Arcas, B.A.: Communication-efficient learning of deep networks from decentralized data. CoRR abs/1602.05629 (2016)
20. Nichol, A., Achiam, J., Schulman, J.: On first-order meta-learning algorithms. CoRR abs/1803.02999 (2018)
21. Pan, S.J., Yang, Q.: A survey on transfer learning. IEEE Trans. Knowl. Data Eng. **22**(10), 1345–1359 (2010)
22. Pennington, J., Socher, R., Manning, C.: GloVe: global vectors for word representation. In: Proceedings of the 2014 Conference on Empirical Methods in Natural Language Processing, pp. 1532–1543 (2014)
23. Rodríguez-Barroso, N., Stipcich, G., Jiménez-López, D., Ruiz-Millán, J.A., Martínez-Cámara, E., González-Seco, G., Luzón, M.V., Veganzones, M.A., Herrera, F.: Federated learning and differential privacy: software tools analysis, the sherpa.ai FL framework and methodological guidelines for preserving data privacy. Inf. Fusion **64**, 270–292 (2020)
24. Smith, V., Chiang, C.K., Sanjabi, M., Talwalkar, A.S.: Federated multi-task learning. In: Advances in Neural Information Processing Systems, vol. 30, pp. 4424–4434. Curran Associates, Inc. (2017)

Use of Artificial Intelligence in Human Resource Management: "Application of Machine Learning Algorithms to an Intelligent Recruitment System"

Said Achchab and Yassine Khallouk Temsamani

Abstract Today, talent and value creation are virtually synonymous. In more and more enterprises, a soaring percentage of market capitalization reflects the direct impact of intangible human capital (Hesketh, A.: Research report valuing your talent CIPD. Chartered Institute of Personnel and Development, 2014). One of the foremost areas the HR should focus upon is "execution". The problem statement revolves around "execution", though the organization has a solid vision, excellent strategy and right talent (Mangipudi, M.R., Vaidya, R.: A study of digitalization in HRM and its effectiveness in execution of HR strategies and policies. Helix 8, 4220–4222, 2018). The HR professionals today are focusing to optimize the combination of human and automated work to gain a simple, seamless, and intuitive work environment. It provides them time for creativity, intelligence, and empathy to deliver an enhanced candidate and employee experience. Richard Coombes, leader of HR transformation practice at Deloitte says that using AI for recruitment eliminates the behavioral and perceptive bias that may happen during human interaction. This article presents an application of artificial intelligence in Recruitment by predicting the likelihood of future success of an applicant.

Keywords HR digitalization · HR tech · Digital recruitment · Predictive hiring · Artificial intelligence · Machine learning

1 Introduction

Now, the importance of Human Resource Function comes from the importance of the human aspect in all types of organizations. The HR Function, as many business and management domains, knew changes and development. The practices of HRM

S. Achchab (✉)
Engineering School of Information Technology and System Analysis, IT Department and Decision Support, University Mohammed V Rabat, Rabat, Morocco

Y. K. Temsamani
Engineering School of Information Technology and System Analysis, Alqualsadi Research Team, University Mohammed V Rabat, Rabat, Morocco

L. Troiano et al. (eds.), *Advances in Deep Learning, Artificial Intelligence and Robotics*, Lecture Notes in Networks and Systems 249,
https://doi.org/10.1007/978-3-030-85365-5_20

were influenced by many factors, such as organizational theory, globalization and the industrial revolutions, including of course the industries 3.0 and 4.0 [1].

All organizations seek profitability, but achieving it means reducing and eliminating all sources of time, effort and money waste. Adopting IT tools and digitally transforming traditional HR methods and processes is a vital key for any organization to produce more while consuming less.

The core of HRM practices are to recruit, develop and retain needed talents for the organization. Adopting digitalization means aiming at adopting a digital recruitment, a digital career development and digital processes and procedure that manage relation and interactions between employees and between employee and organization.

From the above, we can say that the digitalization or digital transformation presents a big opportunity to organizations, especially for enterprises and banks, but what about the threats?

It is true that the digitalization can change HR Function radically, but we cannot talk about digitalization without mentioning Artificial Intelligence. What's the Need for AI in HR?

The best way for an organization to adopt HR digitalization is to start by recruiting. The question to ask here is how to introduce digitalization to recruitment process? What is the role of Artificial Intelligence in recruitment? And what impact may digital recruitment has on HR function and on the organization?

2 Artificial Intelligence

2.1 IA for HR

Today, AI's capabilities are being used to augment business operations and consumer solutions. AI is implanted in HR to: solve pressing business challenges, attract and develop new skills, improve the employee experience, provide strong decision support and use HR budgets as efficiently as possible. The influence of AI is expanding in human resource management. It starts from the recruitment process to exit process of an employee that includes training, engagement, perks, records, and so forth. A few of the influences are highlighted here. Thus, deployment of AI in HR can occur across the entire talent lifecycle (Attract, Hire, Engage, Retain, Develop, Grow, Serve).

There's a lot going on in recruitment right now. Innovative technology is changing the way recruitment works. A main driver behind a lot of the innovation seen in recruitment-land is the use of data, AI (Artificial Intelligence) and predictive analytics (PA).

In recruitment, the potential of AI/predictive analytics is endless: from job advertising and pre-selection to onboarding and funnel optimization, virtually every aspect of recruitment can be streamlined and optimized with an adequate AI/PA-infused recruitment solution [2].

2.2 IA for HR Predictive Analytics (HRPA)

HR analytics is a multidisciplinary approach to integrate methodology for improving the quality of people-related decisions in order to improve individual and organizational performance. There are interchangeable terms used for HR analytics are talent analytics, people analytics, and workforce analytics [3].

Thus, HRPA generates insights that cannot be achieved through traditional benchmarking as HR analytics is reactive and an evidence-based decision system whereas HRPA is proactive and fact-based decision system [4]. Three significant changes that have really created a hunger for predictive analytics in HR [5].

- Major boost in computing power and its affordability.
- HR big data digitally accessible via cloud storage for processing.
- Global talent war to protect and pursue talent streams.

Here are some ways predictive analytics and artificial intelligence can be used during the recruiting process. It can be used to [2].

- Assess which of the qualification requirements were most likely to correlate with good job performance over time.
- Factor in which soft skills will have the most impact on how successful a candidate will be.
- Be more analytical in candidate comparisons.
- Estimate which candidates are more likely to accept a job offer.
- Be more objective in utilizing information gathered during background screening.
- Assess the most successful recruiting methods based on past data.
- More accurately screen out candidates who "look good on paper" but are not likely to do well in the organization.

3 Predictive Hiring and Machine Learning

Recruitment Process

Jean-Marie Peretti [6] presented a recruitment process that contains four major phases as it shown in the following figure.

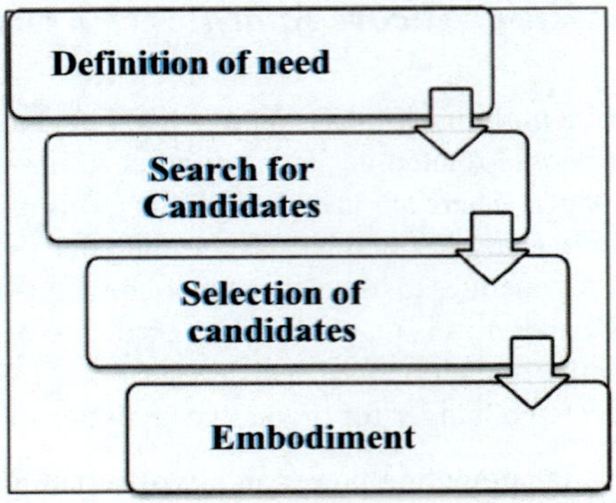

Jean-Marie Peretti, Gestion des Ressources Humaines, Magnard–Vuibert, Juin 2016, pp. 64.

During the first phase of this process, the recruiter receives the request of recruitment (need announcement) from the manager of the department, and then approved, thing that give to the recruiter and the manager green light to start writing the definition of position and also the desired profile.

In the phase of searching, the recruiters choose the method of recruitment (internal or external recruitment) and then proceeding to the 3rd phase that aims at selecting the most suitable profiles for the vacant position. The techniques of selection are various; Peretti [6] mentioned just four:

- 1st selection (pre-selection);
- Survey;
- Interviews;
- And tests.

In the last phase of embodiment, the recruiters alongside with other stakeholders make their final decision (listing the chosen candidates), bargain the contracts with the newcomers and the final step that remain is the integration.

Predictive Hiring

The present article focuses on the selection point to address the issue of predictive hiring and the predictive job performance.

In fact, it can be tough to know which quality is most important when it comes to how successful someone will be. The organization can focus more directly on those characteristics during the recruitment process and perhaps focus less on qualifications that had less of an impact than expected. Organizations can also look at the soft skills that had the greatest impact on future success for an employee and then focus more on finding candidates with those skills. This could even include using the results of personality assessments in a way that helps to predict how successful a given candidate will be.

The recruiting team will be better able to assess combinations of soft skills and experience to see who truly is the best fit and recruiters can be more efficient by focusing on the factors that matter.

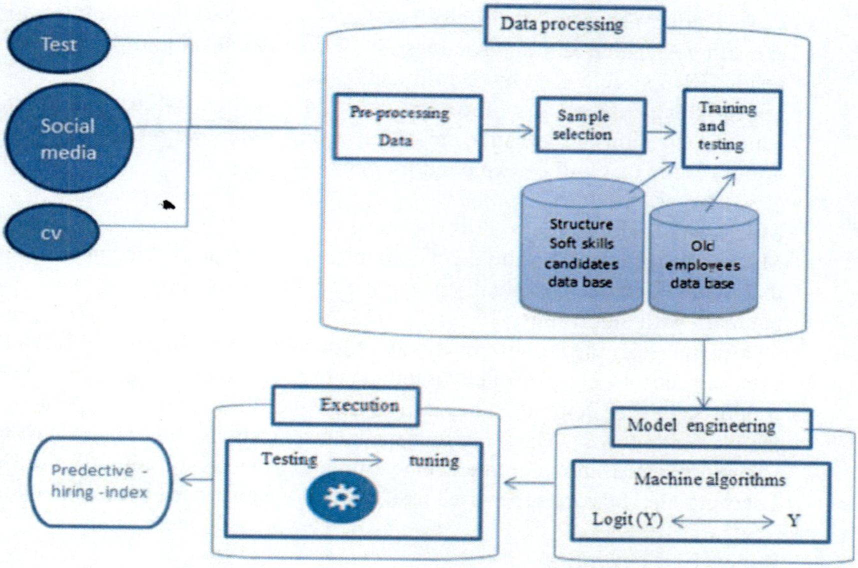

Fig. 1 The predictive hiring model

Predictive analytics for hiring use modern data and assessment science to project candidates' future success based on patterns among current employees. Instead of traditional hiring, which relies on six second resume screens and intuition-based interviews, predictive hiring relies on richer sets of data and smart algorithms to recommend best fit candidates to recruiters and hiring managers.

We propose the model below for predictive hiring (Fig. 1).

First, the data acquisition in this model depends of the results of the candidate's sourcing phase, we propose the use of intelligent questionnaires and semantic search engines so as to search for candidates on professional social networks (LinkedIn, Video, …), recruitment platforms or CV libraries, then extract candidates CVs and classify them while using text mining technologies belonging to artificial intelligence, used when relying on CVs or professional social networks to extract soft skills, for example: IT project manager in a company etc. All these steps will allow us to have a structured database of candidates that we will integrate with the employee's database as an input in this performance Predictive Hiring Analysis.

Secondly, we will use machine learning algorithms to calculate the set of Hiring predictors, our model will be based on five main soft skills (Creativity, Curiosity, empathy, critical, collaboration).

Data acquisition will be pre-processed for sample selection and then data set training take place, in preparation for the execution of routines machine learning.

Next, the model training, where a set of learning data is assigned to the model to make reliable predictions about new or untrained data. At the end, model evaluation, where models are evaluated for performance and effectiveness.

Execution phase in which processed and driven data is transmitted for use in executing Machine learning routines throw Web Application or Web-services calls inside HRIS system used by recruitment team.

And finally, the phase of the deployment in which we will use the techniques of data visualization and real-time model invocation, so that we get a better visualization of our results that will be exploited for decision support.

Machine Learning

Machine learning is the subfield of computer science that gives computers the ability to learn without being explicitly programmed. computers do not learn by reasoning, but learn with algorithms.

Machine learning is one of many subsets of artificial intelligence. When we use one of its subsets, we are in artificial intelligence. Today, there are many ML algorithms proposed in the literature. They can be classified based on the approach used for the learning process. There are four main classifications: supervised, unsupervised, semi-supervised, and reinforcement learning. The two most commonly used Machine Learning methods are supervised learning and unsupervised learning.

Supervised Learning

Supervised learning happens when algorithms are provided with training data and correct answers. The task of the ML algorithm is to learn based on the training data, and to apply the knowledge that was gained using real data, we note that the database must be labeled. There are two types of Supervised Learning techniques: classification and regression.

Classification is the process of predicting a discrete class label or category.

Regression is the process of predicting a continuous value as opposed to predicting a categorical value in Classification.

Unsupervised Learning

ML algorithms do not have a training set (Celebi and Aydin 2016). They are presented with some data about the real world and should learn from that data on their own. Unsupervised learning algorithms are mostly focused on finding hidden patterns in data. In unsupervised learning, we have methods such as clustering.

Clustering is one of the most popular unsupervised machine learning techniques used for grouping data points or objects that are somehow similar, it is used mostly for: Discovering structure, Summarization and Anomaly detection (Fig. 2).

4 The Machine Learning Model

The key to not getting lost in the huge space of machine learning algorithms is to realize that it consists of combinations of just three component Learning = Representation + Evaluation + Optimization [8].

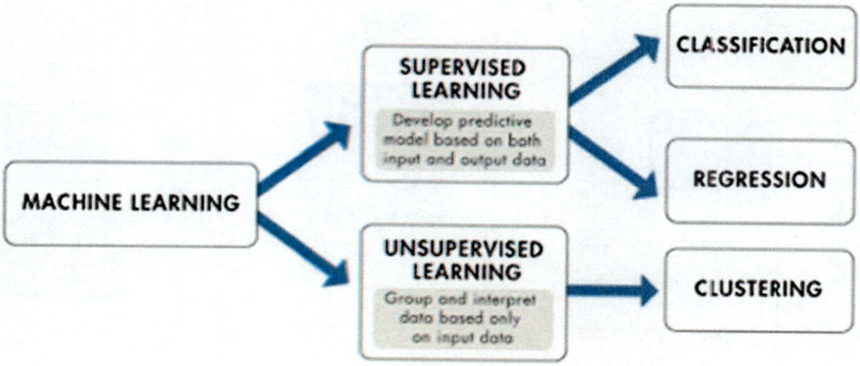

Fig. 2 Supervised learning versus unsupervised learning [7]

4.1 Pre-processing

Before creating a model, we divide our dataset into training (80%) and testing (20%).

We divided this data set into training because this is the data set that we will be feeding to the logistic regression model, this is how we will train the model.

When the model is created, we will test that model using the testing data set to validate it.

4.2 Representation of the Model

In this article we will predict if the candidate will perform good in the entity using tree classification algorithms: Logistic Regression, KNN and SVM (Fig. 3).

5 Representation of ml Algorithms of the Models

Logistic Regression (LG)

Logistic regression is a method for binary classification. It works to divide points in a dataset into two distinct classes, or categories. For simplicity, let's call them class A and class B. The model will give us the probability that a given point belongs in category B. If it is low (lower than 50%), then we classify it in category A. Otherwise, it falls in class B. It's also important to note that logistic regression is better for this purpose than linear regression with a threshold because the threshold would have to be manually set, which is not feasible. Logistic regression will instead create a sort of S-curve (using the sigmoid function) which will also help show certainty, since the output from logistic regression is not just a one or zero. Here is the standard logistic

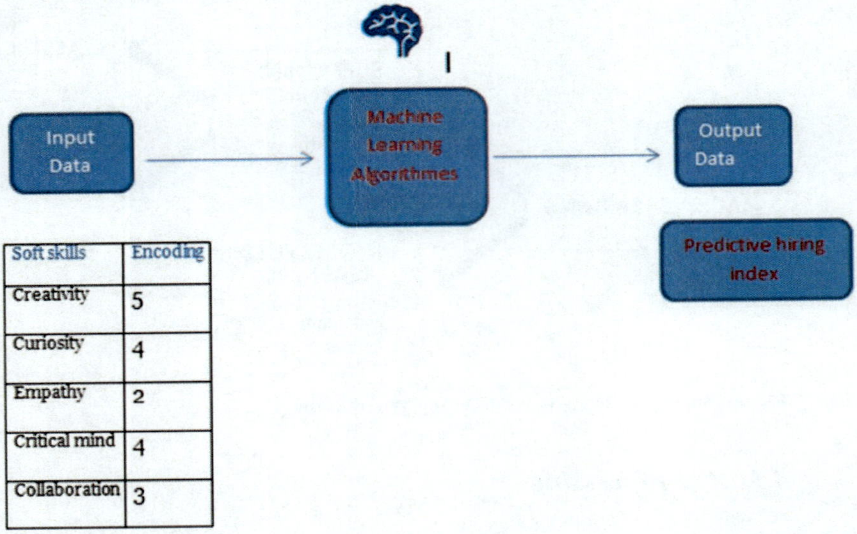

Fig. 3 The basics of machine learning technology

function, note that the output is always between 0 and 1, but never reaches either of those values.

- Sigmoid Function (Logistic Function)

Logistic regression algorithm also uses a linear equation with independent predictors to predict a value. The predicted value can be anywhere between negative infinity to positive infinity. We need the output of the algorithm to be class variable, i.e. 0—no, 1—yes. Therefore, we are squashing the output of the linear equation into a range of (0–1). To squash the predicted value between 0 and 1, we use the sigmoid function [python machine learning by AMIRSINA TORFI].

$$z = \theta_0 + \theta_1 \cdot x_1 + \theta \cdot x_2 + \cdots \qquad g(x) = \frac{1}{1+e^{-x}}$$

$$h = g(z) = \frac{1}{1+e^{-z}}$$

We take the output (z) of the linear equation and give to the function g(x) which returns a squashed value h, the value h will lie in the range of 0–1. To understand how sigmoid function squashes the values within the range, let's visualize the graph of the sigmoid function.

- Cost Function:

Since we are trying to predict class values, we cannot use the same cost function used in linear regression algorithm. Therefore, we use a logarithmic loss function to calculate the cost for misclassifying [towards data scientist by Rohith Gandhi].

- Log loss (logarithmic loss function):

Measures the performance of a classification model where the prediction input is a probability value between 0 and 1. The goal of our machine learning models is to minimize this value.

K-Nearest Neighbor (K-NN)
KNN is the simplest machine learning algorithm used for classification and regression. It makes decision based on the entire training dataset. No time is spent on training the algorithm but only for pre-processing and testing. The algorithm makes predictions by calculating similarity between the input sample and each training instance. This algorithm does not make strong assumptions about the form of mapping function hence it is Nonparametric. In simple words, by not making assumptions, the algorithm is free to learn any functional form from the training data.

K represents the number of training data points lying in proximity to the test data point which we are going to use to find the class.

Support Vector Machine (SVM)
A Support Vector Machine (SVM) is a very powerful and versatile Machine Learning model, capable of performing linear or nonlinear classification, regression, and even outlier detection. It is one of the most popular models in Machine Learning, and anyone interested in Machine Learning should have it in their toolbox. SVMs are particularly well suited for classification of complex but small- or medium-sized datasets [9].

6 Evaluation of Model

The Confusion Matrix [10]
The aim is to produce a model to predict as precisely as possible the values taken by a categorical variable Y. A preferred approach to evaluate the quality of the model would be to compare the predicted values with the true values taken by Y: it's the role of the confusion matrix (Fig. 4).

The entries in the confusion matrix have the following meaning in the context of our study:

- **a** and **d** are respectively the true Negative and Positive prediction
- **c** and **b** are respectively the False Negative and Positive prediction

Fig. 4 The confusion matrix

		Predicted	
		Negative	Positive
Actual	Negative	a	b
	Positive	c	d

Accuracy
The accuracy (AC) is the proportion of the total number of predictions that were correct [10]. It is determined using the equation:

$$\text{Accuracy} = \text{number of correct prediction/total number}$$

With an accuracy of 80% we can deduce that the model proposes correct predictions.

The recall or true positive rate.

The recall or true positive rate (TP) is the proportion of positive cases that were correctly identified [10].

The false positive rate (FP) which is the proportion of negatives cases that were incorrectly classified as positive,

The true negative rate (TN) and t**he false negative rate** (FN).

Precision
Precision (P) is the proportion of the predicted positive cases that were correct [10].

$$Precision = \frac{TruePositives}{TruePositives + FalsePositive}$$

Roc Curve
AUC ROC is one of the most important evaluation metrics for any classification model's performance. ROC (Receiver Operating Characteristic) Curve tells us about how good the model can distinguish between two things. Better models can accurately distinguish between the two. The best possible AUC is 1 while the worst is 0.5 (the 45° random line). Any value less than 0.5 means we can simply do the exact opposite of what the model recommends getting the value back above 0.5 [11].

F1-score
F1-score, is simply the harmonic mean of precision and recall [11].

$$\text{F1 Score} = 2 * \frac{\text{Precision} * \text{Recall}}{\text{Precision} + \text{Recall}}$$

Mean Square Error (MSE)

The square of the difference between the predicted and actual target variables, divided by the number of data points Mathematically, it is represented as [9].

$$\mathrm{MSE} = \frac{1}{n}\sum_{i=1}^{n}(y_i - \tilde{y}_i)^2$$

In the literature we find also Mean Absolute Error (MAE) and the Matthews correlation coefficient (MCC).

7 Optimization

7.1 *Regularization*

Regularization is a technique used to avoid the overfitting problem. Overfitting happens when a model learns the detail and noise in the training data to the extent that it negatively impacts the performance of the model on new data [9].

7.2 *Gradient Descent*

Gradient descent is a first-order iterative optimization algorithm for finding the minimum of a function. The general idea of Gradient Descent is to tweak parameters iteratively to minimize a cost function [9].

8 Simulation

The results of simulation of our model with the three ML algorithms are summarized in Table 1.

All the algorithms perform good but the SVM is most perfect one for our case. What can be deduced from this summary table, that with an accuracy of 95%, a very low MSE, a recall and a high precision and a small range between the accuracy of training and testing, we can conclude that the model performs perfectly and we were able to fight against overfitting.

Table 1 Performance evaluation of the three models

Metrics	KNN	Logistic regression		SVM
		Avant l'optimisation	Apres l'optimisation	
Accuracy-test Accuracy-train	0.95(95%) 0.83(83%)	0.85(85%) 0.83(83%)	0.95(95%) 0.85(85%)	0.95(95%) 0.92(92%)
Recall	0.94	0.83	0.95	0.944
Precision	1.0	1.0	1.0	1.0
MSE	0.05	0.15	0.05	0.1
MAE	0.05	0.15	0.05	0.1
Fl-score	0.95	0.87	0.95	0.954
AUC ROC curve	0.972	0.91	0.972	0.972
Correlation	0.793	0.577	0.69	0.80

9 Conclusion

In this article the three **algorithms** we treated give good results and are similar. Nevertheless, in our case, the SVM seems to be the best one due to the smallest error value between training and testing. AI-based HR applications raise employee productivity. It has the ability to analyse, predict, diagnose and become a more capable resource while focusing on employee needs and outcomes.

However, there are challenges like privacy, talent gap, maintenance, integration capabilities, or limited proven applications. But exercising proper caution while deploying AI services will solve unnecessary issues. AI systems must be managed carefully by finding reliable learning data sets, using the right implementation approach, seeking clarity, eliminating bias, and considering unintentional consequences [12].

References

1 Shamim, S., Cang, S., Yu, H., Li, Y.: Management approaches for Industry 4.0: a human resource management perspective (2016)
2 Neelie: An introduction to predictive analytics in recruitment, February 12, 2018. https://harver.com/blog/hr-tech-trend-an-introduction-to-predictive-analytics/
3 Mishra, S.N., Lama, D.R., Pal, Y.: Human resource predictive analytics (HRPA) for HR management in organizations. Int. J. Sci. Technol. Res. **5**(05) (2016)
4 Handa, D., Garima: Human resource (HR) analytics: emerging trend in HRM (HRM). IJRCM **5**(06), ISSN: 0976–2183 (2014)
5 Ladimeji, K.: 5 things that HR predictive analytics will actually predict. Recruiter sec. I, 1 (2013)
6 Peretti, J.-M.: le recrutement selon Jean-Marie Peretti, Vuibert (2016)
7 Bunker, R.P., Thabtah, F.: A machine learning framework for sport result prediction (2019)
8 Domingos, P.: A few useful things to know about machine learning (2012)

9 Géron, A.: Hands-On Machine Learning with Scikit-Learn and TensorFlow: Concepts, Tools, and Techniques to Build Intelligent Systems (2017)
10 Kohavi, R., Provost, F.: Glossary of terms. In: Machine Learning **30**(2–3) (1998)
11 Fawcett, T.: An introduction to ROC analysis. Pattern Recogn. Lett. **27**(8), 861–874 (2006)
12 Christopher, A.: Use of artificial intelligence in human resource management. In: ACM Woodstock Conference WOODSTOCK'18, ISBN: 978-1-4503-0000-0/18/06. El Paso, Texas USA (2018)
13. Hesketh, A.: Research report valuing your talent CIPD. Chartered Institute of Personnel and Development (2014)
14. Mangipudi, M.R., Vaidya, R.: A study of digitalization in HRM and its effectiveness in execution of HR strategies and policies. Helix **8**(6), 4220–4222 (2018)

Combining Active Semi-supervised Learning and Rare Category Detection

Rohan Loveland and Noah Kaplan

Abstract The task of exploring data sets that are too large to allow individual examination of all instances necessitates new tools to identify and group interesting classes. This work investigates a hybrid system that performs both rare category detection and active semi-supervised clustering using the previously developed Farpoint algorithm. The prior algorithm focused solely on rare category detection, while the current research incorporates a capability for simultaneously obtaining high clustering accuracy. We investigate two methods of constrained clustering. One method is a constrained variant of K-Means that favors compact clusters, while the other is a novel variant of hierarchical agglomerative clustering that determines a cut-off threshold based on constraint violation. Using artificially skewed MNIST data we empirically compare the relative performance of the two new methods with the original Farpoint algorithm. Results show substantial improvement in clustering accuracy.

Keywords Semi-supervised learning · Rare category detection · Constrained clustering

1 Introduction

The problem of "*rare category detection*" (RCD) is formally defined as: "Given a set of unlabeled examples $S = \{x_1, \ldots, x_n\}$, $x_i \; in\mathbb{R}^d$, which come from m distinct classes, i.e. $y_i \in 1, \ldots, m$, the goal is to find at least one example from each class by requesting as few total labels as possible" [5]. Semi-supervised learning is similar

R. Loveland (✉)
Augustana College, Rock Island, IL, USA

N. Kaplan
Whitman College, Walla Walla, WA, USA
e-mail: kaplannp@whitman.edu

L. Troiano et al. (eds.), *Advances in Deep Learning, Artificial Intelligence and Robotics*, Lecture Notes in Networks and Systems 249,
https://doi.org/10.1007/978-3-030-85365-5_21

in that it uses unlabeled data together with labeled data, but with a different goal, namely, building better classifiers/clusterers [16]. The previous Farpoint algorithm was developed strictly for performing RCD, without considering clustering accuracy [8]. In practice both RCD and accurate clustering are desirable in realistic data exploration. For example, if using RCD on an astronomical data set revealed the existence of a UFO, it's likely that it would then be desirable to identify the other UFOs using a concurrently developed semi-supervised clusterer. In this paper we modify the previous Farpoint algorithm to create a hybrid algorithm that provides good clustering accuracy without sacrificing RCD performance.

In order to make this change *constrained clustering* is integrated into the original algorithm, which split clusters by assigning unlabeled points to the nearest labeled (i.e. queried) point. In the following we compare the original algorithm, here designated as Farpoint-NQ, named for *"Nearest Query"* with two other variants, designated Farpoint-COPK and Farpoint-HA. These are named, respectively, for the *"Constrained K-Means"* and *"Hierarchical Agglomerative"* constrained clustering algorithms [13].

We evaluate the algorithm's performance on artificially skewed versions of the MNIST data set [7]. This allows controlled investigation of differences in performance due to both the effects of differing relative frequency of representation of classes ("class probabilities") and inherent class structure differences in feature space.

2 Related Work

Here we provide a summary of previous work, and a review of the previous Farpoint algorithm.

2.1 Previous Work

The structure of data is used to find rare categories. Previous examples include using Eigenbasis Modeling of Uninteresting Data (DEMUD), which is based on principal components modeling and reconstruction error [15]. DBSCAN [1] uses a similar approach based on density based clustering. In many cases, however, it is also possible to get information from a user (oracle). When this information is available, it can be used to improve RCD performance and to find interesting anomalies [10].

Hospedales [6] points out that "Discovering rare categories and classifying new instances of them is an important data mining issue in many fields, but fully supervised learning of a rare class classifier is prohibitively costly in labeling effort. There has therefore been increasing interest both in active discovery: to identify new classes quickly, and active learning: to train classifiers with minimal supervision." The new Farpoint algorithm accomplishes this by use of semi-supervised clustering algorithms.

Active learning as a technique to provide more labeled samples as input for training a classifier has been criticized as potentially ineffective in some circumstances [9]. However, Wagstaff et al. have continued to explore and develop this technique successfully, particularly for imagery datasets [14].

Toon Von Kraenendonck provides a summary of current state of the art semi-supervised clustering algorithms and presents a model selection algorithm for semi-supervised clustering, COBRAS [12]. We use one of the algorithms he discusses, COPKMeans [13], and a custom version of constrained Hierarchical Agglomeration.

2.2 Previous Farpoint Algorithm

The Farpoint-NQ algorithm is an active learning semi-supervised clustering algorithm, based on the hierarchical division of an initially unlabeled data set into a partition of increasingly refined clusters [8].

Assume we're given a set of n data points $X = \{x_1, \ldots, x_n\}$, $x_i \in \mathbb{R}^d$, $i \in \{1, \ldots, n\}$. These points come from m distinct classes, with corresponding labels $y_i \in \{1, \ldots, m\}$ that are initially unknown. A label is obtained using a query, which we will designate as query function $QRY(i)$, where $QRY(i) = y_i$. $QRY(i)$ represents the process of presenting the user with the data sample with index i and asking for a label; the resulting answer is assumed to always be correct.

Farpoint begins with a partition $P_0 = \{C_0\}$, where initially $C_0 = X$, the set of all data points. As the algorithm continues and more queries are asked the partition is refined, by splitting existing clusters. Then $P_k = \{C_0, C_1, \ldots, C_j\}$, where $j \leq k$, where k is the number of queries.

Initially the index of a point near the edge of the data set, q_0, is determined. C_0 is then supplied a label by querying the label of this point, resulting in cluster label $l_0 = QRY(q_0)$. In general q_k is the index into X of the most recent query, with $q_k \in \{1, \ldots, n\}$, and l_j is the label of C_j. C_0, and more generally any cluster C_j, can be seen as the union of two disjoint sub-clusters, such that $C_j = I_j \cup O_j$. I_j contains points with known labels (obtained through queries), and O_j contains points without known labels that are nonetheless assumed to be part of C_j. Initially, I_0 contains only the starting point, so that $I_0 = \{x_{q_0}\}$. All the other points are assigned to O_0, so $O_0 = \{x_i | \forall i, i \neq q_0\}$.

Then assume an algorithm $FP(C_j)$ which returns the index of the cluster's *far point* fp_j, which is the point in O_j that is most distant from all the points in I_j. Mathematically, $fp_j \in O_j$ has the property that the distance $\min\{|fp_j - x_a|\} \geq \min\{|x_b - x_a|\}, \forall x_a \in I_j, \forall x_b \in O_j$.

(The details of the $FP()$ algorithm are discussed below.) The far point is assumed to be the point that is least likely to belong to the current cluster, and so it's used for determining next query of cluster C_j.

$FP(C_0)$ is then used to find which point in O_0 to query using $q_1 = FP(C_0)$. The corresponding label for x_{q_1} is then determined using the fact that $QRY(q_1) = y_{q_1}$.

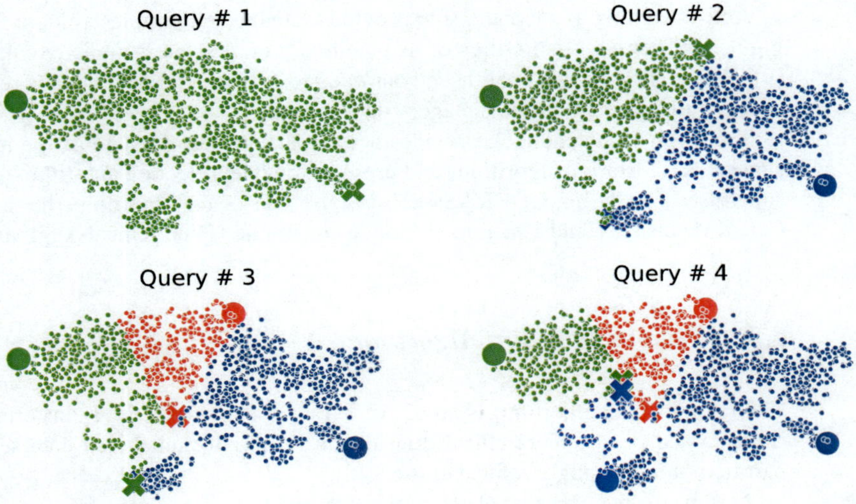

Fig. 1 Far point-NQ: in this figure the MNIST data has been reduced using tSNE from 784 to 2 dimensions, in order to allow visualization. The top left image shows the initial cluster with all points assumed to belong to a single label; the others show FP-NQ's refinement of the partition as more queries are made. In all images the digit class is indicated by color, the already queried points are shown with dots, and the far points are shown with x's

If this label is the same as the label of the cluster, such that $y_{q_1} = l_0$, then this newly queried point has been verified to belong in the current cluster. I_0 and O_0 can then be updated such that $I_0 \leftarrow I_0 \cup \{x_{q_1}\}$, while $O_0 \leftarrow O_0 \setminus \{x_{q_1}\}$.

Alternatively, if $y_{q_1} \neq l_0$, then this point belongs in a new cluster, and C_0 will need to be split. More generally, if the newly queried far point's label doesn't match the cluster's current label, such that $y_{q_k} \neq l_j$, current cluster C_j will need to be split. This is accomplished in Farpoint-NQ by creating a new cluster, C_{j+1}, with corresponding I_{j+1} initially set to just $\{x_{q_k}\}$. The points in O_j are then divided between O_j and O_{j+1}, based on whichever previously queried point in I_j or I_{j+1} is nearest.

Once the query has been processed, the partition P_0 is then updated if a cluster has been newly created, in which case that $P_1 = \{C_0, C_1\}$. Otherwise, $P_1 = \{C_0\}$, but C_0 will have a new fp_0 to query.

The querying process then continues. In order to select the cluster to query next, a function $FPD(C_j)$ is used to calculate the distances from each cluster's fp_j to the closest point in the corresponding I_j. The cluster with the largest $FPD(C_j)$ is then selected. The particular point within this cluster to query is found using index $FP(C_j)$, and the cluster is then processed as described previously.

The query process continues for as many labels as the user decides to provide. Figure 1 shows the basic flow of the algorithm. The algorithm is also described in pseudo-code below as Algorithm 1.

Algorithm 1 Far Point-NQ

Require: X: an unlabeled set of data points
P_k: a partition holding a list of clusters
C_j: a cluster object, containing:
– l_j—the label of cluster C_j
– I_j: a set of data points assigned to C_j with known labels—$\{x_r | x_r$ have been queried previously$\}$
– O_j: a set of data points without known labels—$\{x_s | x_s$ have not been previously queried$\}$
– fp_j: a *far point* in C_j, $fp_j \in \{O_j\}$
– $QRY(i)$—a function to get y_i, the label of point x_i
– $FP(C_j)$—a function to get the index into X of far point fp_j
– $FPD(C_j)$—a function to get the largest distance from fp_j to the known points in C_j

Ensure: Attempt to minimize number queries while finding all classes in X
form initial partition P_0 and cluster $C_0 = X$
find q_0, the index of a point on the edge of X, and set $l_0 = QRY(q_0)$
initialize I_0, O_0, and calculating $fp_0 = FP(C_0)$
while user continues search for more classes **do**
  find index j^* of next cluster to query by selecting cluster with greatest $FPD(C_j)$
  perform query, setting $l_{new} = QRY(FP(j^*))$
  if new query label equals cluster label, i.e. $l_{new} = l_{j^*}$ **then**
    assign fp_{j^*} to I_{j^*}, removing from O_{j^*}
    recalculate fp_{j^*} using $fp_{j^*} = FP(C_0)$
  else if new query label is not equal to cluster label, i.e. $l_{new} \neq l_{j^*}$ **then**
    split C_{j^*} into two, creating C_{j^*+1}
    assign fp_{j^*} to I_{j^*+1}
    reassign O_j^*, O_{j^*+1} pts based on closest pts in I_{j^*}, I_{j^*+1}
    update far points in both clusters fp_{j^*}, fp_{j^*+1}
  end if
end while

The previous description relies on the existence of algorithms $FP(C_j)$, which is used to calculate the index of fp_j. The procedure for finding this is described in pseudo-code below as Algorithm 2.

3 Algorithms

The previous algorithm, FP-NQ, is targeted strictly towards RCD. This is reflected in the splitting strategy. When cluster C_j is split due to a query of fp_j returning a different label than the current cluster (i.e. $y_{q_k} \neq l_j$), the assignment of the points belonging to the current cluster that are unverified, O_j, is reevaluated. In the FP-NQ algorithm, they are simply assigned to whichever's closest of either the previously labeled points in I_j or fp_j.

This splitting strategy takes no account of the structure of the points in cluster C_j. In the event that the far point fp_j is part of a different, clearly separated sub-cluster, FP-NQ will not recognize this, and will instead continue by evenly dividing

Algorithm 2 Algorithm to find cluster far point-$FP(C_j)$

Require: C_j: a cluster with:
- I_j: a set of data points assigned to C_j with known labels—$\{x_r | x_r$ have been queried previously$\}$
- O_j: a set of data points without known labels—$\{x_s | x_s$ have not been previously queried$\}$

Ensure: return f: an index into X that ensures distance $\min\{|x_f - x_a|\} \geq \min\{|x_b - x_a|\}, \forall x_a \in I_j, \forall x_b \in O_j$

1: **for** each $x_b \in O_j$ **do**
2: store distances $|x_b - x_a| \forall x_a \in I_j$ into a column vector
3: concatenate columns into a table
4: **end for**
5: **for** each column in table **do**
6: find minimum value
7: store that minimum in a row vector
8: **end for**
9: set f = index of maximum value in row vector

(approximately) the points from O_j, thus mis-assigning a significant number of them to the wrong cluster. This may not affect RCD performance, but it can have a significant effect on cluster accuracy.

It is thus possible to preserve the RCD performance of the Farpoint algorithm but improve the clustering accuracy by incorporating a clustering algorithm during the splitting process. The presence of user-supplied labels requires the usage of *constrained* clustering algorithms, because the *must-link* (ML) and *cannot-link* (CL) constraints implied by the labels must be conserved.

Generally, types of clustering include centroid-based and connectivity-based. We try one of each type—COPK and HA, at the splitting step of Farpoint.

COPK is centroid-based, and therefore favors spherical clusters. HA is connectivity-based, which allows elongated clusters. We expect HA to have better performance in general, given the frequent occurrence of non-spherical class boundaries in real-world data sets.

3.1 Farpoint-NQ

We use the basic Farpoint algorithm to provide a baseline to assess the performance improvements in classification of the COPK and HA clustering algorithms. The Farpoint-NQ strategy of unlabeled point assignment is not a clustering algorithm in the normal sense, in that no account of underlying connectivity or density is taken into account.

3.2 Farpoint-COPK

COPK is a constrained K-means algorithm, which utilizes pairwise must-link and cannot-link constraints incorporated into the normal K-Means iterative structure. Specifically, for a given iteration, it assigns each data point to the nearest existing cluster centroid such that the assignment doesn't violate any of the constraints [11, 13]. In Farpoint-COPK each of the label points as well as the far point of the cluster are used as initial seeds.

3.3 Farpoint-HA

Connectivity-based clustering algorithms allow elongated structure (i.e. non-spherical) to be found. Some methods, such as DBSCAN and Mean-shift, require a parameter like ϵ or a kernel radius that must be specified in advance [2, 3]. Such a parameter must be properly scaled in a way that is data set dependent, and can be difficult for a user to determine. This is related to the *model selection problem*, where the number of clusters must be specified, in, e.g. K-Means [4].

In order to avoid this we propose using a hierarchical agglomerative (HA) clustering scheme. HA methods allow for the number of clusters to be determined either explicitly, or by specifying a cut-off threshold for the maximum distance between clusters. That inter-cluster distance can have several possible definitions; we use the *single-linkage* distance, which is the minimum distance between any of the points in one cluster and any of the points in the other, $= \min\{d(x_a, x_b)|x_a \in C_a, x_b \in C_b\}$.

The specification of a clustering parameter can be avoided by allowing the cut-off threshold to be the largest possible value that does not violate a constraint.

This process is illustrated with a synthetic data set in Fig. 2. On the left of the figure we have a total of eight 2-D data points. All are originally assumed to belong to the same cluster C_j that is about to be queried. We can assume the blue point on the far left is labeled, so $\in I_j$, and this means the orange point on the right is C_j's farpoint fp_j. The query will result in a different label for fp_j that doesn't match the cluster label l_j, so the cluster will be split, with all of the other blue and orange points, currently in O_j, needing to be assigned to a cluster.

The dendrogram shows that both of the orange points will be clustered, as well as all of the blue points except for the outlier. The outlier is orphaned because the constraint violation cut-off is hit before it can be assigned. The cut-off is present because we know the orange cluster containing the farpoint (on the right) has a CL constraint that prevents it from being joined with the blue cluster on the left.

Once this constraint cut-off threshold has been reached, no further HA clustering is possible without violating the constraints. This can leave some *orphans*, however, that are too far away to be assigned to either cluster by HA. In the figure this is the blue outlier. In this case we assign the orphan to whichever cluster it's closest to, by the single link distance criterion.

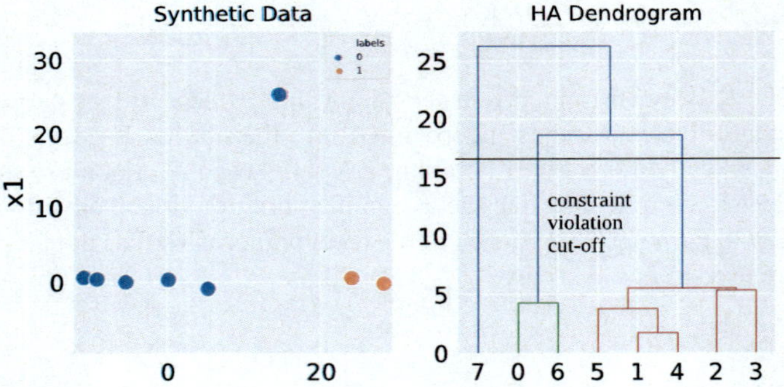

Fig. 2 Far point-HA: in this figure we illustrate the splitting of a single cluster of a set of points on the left with actual class labels shown in blue and orange, and a dendrogram on the right showing the cut-off threshold determined by the corresponding constraint violation

4 Experimental Evaluation

In order to obtain a data set with rare categories, the MNIST data set was artificially skewed over multiple trials, changing the probability of each class. For a given trial, the data set was randomly sampled over the 10 classes related approximately by powers of 2.

The relative frequency of the most frequently represented class to the least is about 500:1, providing a broad variation to work with. Classes that have a lot in common (e.g. 4's, 9's in MNIST) have similar structure that pose challenges beyond sparsity. It's desirable to examine each class at all relative frequency levels, in order to understand the effects of both sparsity and structure within the feature space.

Covering the entire space of all class/probability representations is prohibitively computationally expensive; there are approximately 3.6 million (i.e. 10!) different possibilities. Given 1000 runs selected at random however, each (class/probability) pair is present in about 100 runs, which has previously been shown empirically given a random query selection strategy to be sufficient [8].

The FP-NQ and FP-COPK algorithms were accordingly run 1000 times each, while HA was run 200 times. This is due to the fact that the FP-HA algorithm is significantly more computationally expensive. In brief, the time complexity of a naive agglomerative clustering is $O(n^3)$ because of exhaustively scanning the N × N distance matrix for the lowest distance in each of N − 1 iterations.

In some cases the algorithms overflow, taking more than the allotted maximum 1000 queries to find the sparsest class. Specifically,we observed approximately 13% for FP-NQ, 9% for FP-COPK, and 3% for FP-HA. Given the underlying stochastic nature of the algorithms, this is likely the result of a poor seed initialization, and can be re-run.

The time complexity of the algorithms varies by roughly a factor of three between NA and COPK, and then by roughly 1000 for HA. This is summarized in Table 1

Table 1 Average computation times per trial

Ave. time (s)			
Algorithm	NQ	COPK	HA
Finding 10 classes	20	72	53,000
1000 queries	60	107	70,000

The run times given are approximate, since the computations were run on and then aggregated from different computers.

While FP-NQ and FP-COPK run in minutes, FP-HA takes hours.

Since we are interested in both the RCD and clustering accuracy, we compare the performance of all three algorithms in both.

5 Performance

5.1 Rare Category Detection

The goal of RCD is to find all the classes with as few queries as possible. The performance of each of the methods, on a per digit basis, is shown in Figs. 3, and 4.

Generally, we expect the number of queries required to find a class to be higher for classes that are sparsely represented (i.e. with lower class probability). This is clearly borne out in all three plots.

The difficulty of finding the different digit classes clearly varies. In all three plots the same order of difficulty is maintained for five of the ten classes. The two classes that are most difficult, one's followed by nine's, are seen at the top of each plot. The three least difficult are seen similarly at the bottom of the plots: zero's followed by two's and then six's.

One's are in general the hardest to find, and also have the highest variance in all 3 plots. We believe that this is because the shape of the numeral '1', with a single centered vertical stroke, is most similar to all the other digits. We hypothesize the variance is due to the dependence on the level of representation of other classes.

The nine's are next in order difficulty; this is likely because the numeral '9' can be very similar in appearance to '4'.

Conversely, zero's are consistently easiest to find. This is likely because the numeral '0' has nothing in the middle, thereby sharing fewer pixel locations with the other digits.

The consistent large variation in the number of queries required to find the various digit classes makes it clear that *structure*, as well as sparsity, plays a key role in determining the required number of queries to find a class.

A comparison of the overall RCD performance of the three algorithms averaged over all digits is shown in Fig. 5. This makes it clear that all FP-NQ, FP-COPK,

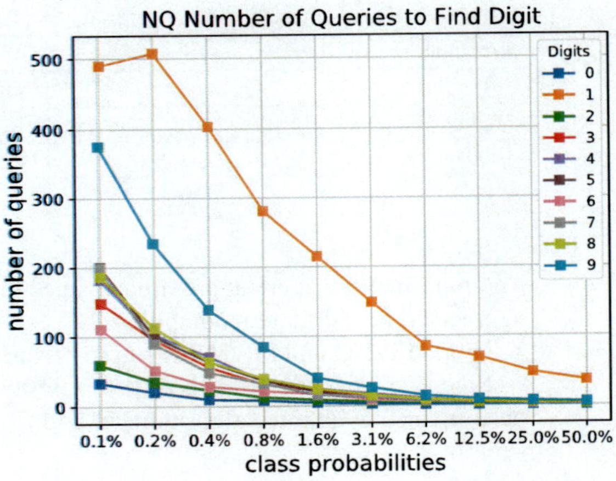

Fig. 3 Farpoint-NQ: rare category detection performance in terms of number of queries by class probabilities, shown separately for each digit

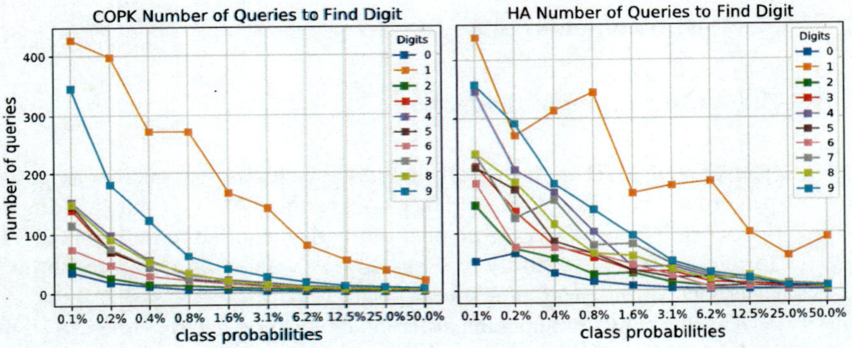

Fig. 4 Farpoint-rare category detection in terms of number of queries by class probabilities, shown separately for each digit. (left) Farpoint-COPK (right) farpoint-HA

and FP-HA all beat a random query strategy by a wide margin. The slopes of the curves also indicate that this performance gain will become even more pronounced as classes become sparser, with potential orders of magnitude improvement.

Between the three algorithms COPK proves better than NQ by a narrow margin, while both have a wide margin over HA (Table 2).

5.2 *Clustering Accuracy*

The goal of semi-supervised learning is to utilize the structure of unlabeled data to improve classification/clustering accuracy. The performance of each of the methods, on a class probability basis, is shown in Figs. 6 and 7. Here the digit performances

Table 2 Average number of queries for far point alg. and random strategy: performance of the random strategy is shown compared to the three far point alg. variants

Class frequencies										
%	0.1	0.2	0.4	0.8	1.6	3.1	6	12	25	50
Rdm	1K	500	250	125	63	31	16	8	4	2
NQ	182	127	87	59	40	27	15	14	8	6
CPK	156	100	68	54	36	25	17	12	8	5
HA	256	164	132	96	59	44	31	23	10	11

This is graphed in Fig. 5

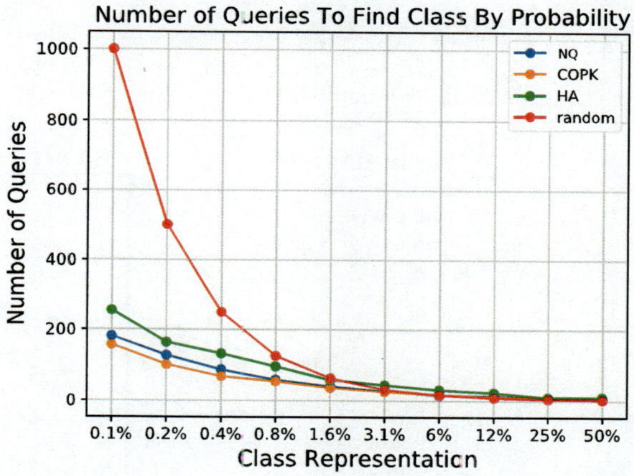

Fig. 5 Farpoint-NQ, COPK, and HA rare category detection performance: number of queries as a function of class probability, averaged over all digits

have been averaged out, and the categories shown are the class probabilities. The plots show similar behavior in all three cases in that the most populated classes have the highest accuracies.

It is interesting that NQ, the simplest method, appears to have initially higher clustering accuracy for low numbers of queries. This is likely because 50% clustering accuracy is easily attained by classifying the entire dataset as one class. In the short term, the more sophisticated clustering strategies of COPK and HA give greater weight to sparser classes, leading to worse performance, but these differences are made up as the querying progresses. This is more clearly demonstrated in Fig. 8, which shows the results averaged over all digits.

Figure 8 shows the accuracy averaged over the class probabilities. In this case we see that FP-NQ and FP-COPK have very similar performance, while FP-HA is the clear winner past 300 queries. At 1000 queries the accuracies are: NQ—82.5%, COPK—84.9%, and HA—95.9%. This represents nearly a 4× improvement in terms of reduced losses for FP-HA over FP-COPK.

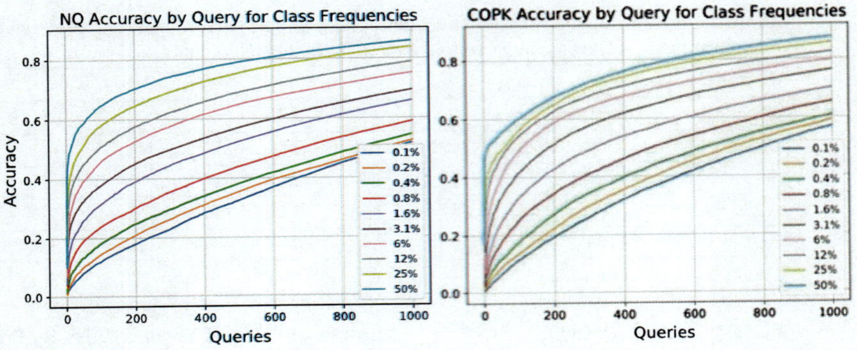

Fig. 6 Farpoint—clustering accuracy by queries, with each digit shown separately, averaged over 1000 trials. Farpoint-COPK (left), farpoint-NQ (right)

Fig. 7 Farpoint-HA: clustering accuracy by queries, with each digit shown separately, averaged over 200 runs

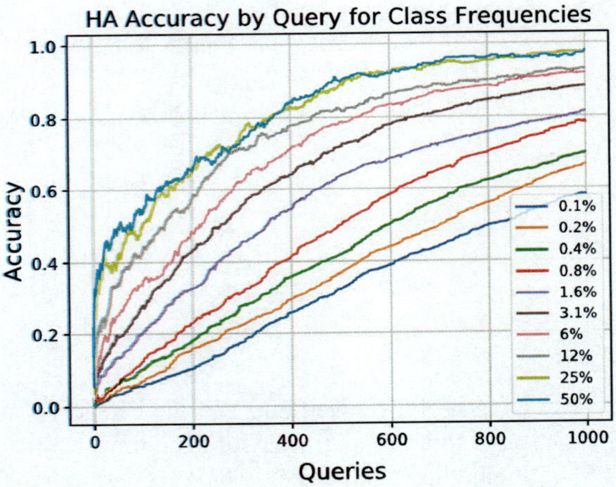

Fig. 8 Farpoint-NQ, COPK, and HA: clustering accuracy by queries, averaged over all digits

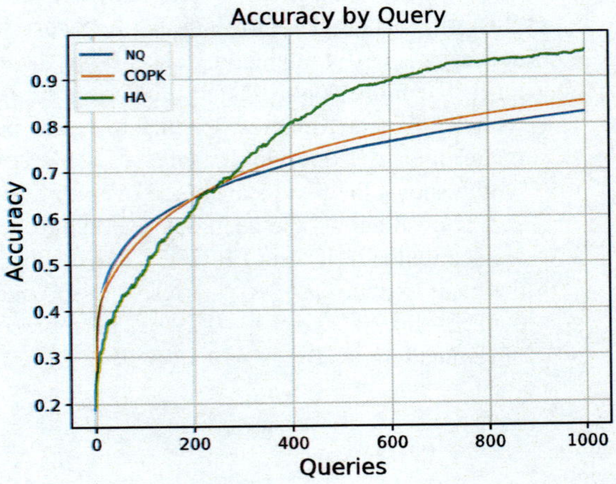

6 Conclusions

It's clear that a hybrid algorithm can be developed which performs both RCD and accurate clustering. FP-COPK is narrowly better than FP-NQ for both, with a roughly 2× computation time increase, which might be trivial in many circumstances. FP-HA is somewhat worse at RCD, but provides dramatically better accuracy performance. Unfortunately, the time complexity increase is substantial, so it may be impractical for many applications without streamlining the algorithm. The results also make it clear that in all cases both sparsity and *structure* matter significantly. Experiments that only investigate sparsity do not reveal the entire story. All three algorithms perform significantly better than a random query selection strategy, with potential order of magnitude improvements for large skew values.

References

1. Chen, Z., Li, Y.F.: Anomaly detection based on enhanced DBScan algorithm. Procedia Eng. **15**, 178–182 (2011)
2. Cheng, Y.: Mean shift, mode seeking, and clustering. IEEE Trans. Pattern Anal. Mach. Intell. **17**(8), 790–799 (1995)
3. Ester, M., Kriegel, H.P., Sander, J., Xu, X., et al.: A density-based algorithm for discovering clusters in large spatial databases with noise. In: KDD, vol. 96, pp. 226–231 (1996)
4. Forgy, E.W.: Cluster analysis of multivariate data: efficiency versus interpretability of classifications. Biometrics **21**, 768–769 (1965)
5. He, J., Carbonell, J.G.: Nearest-neighbor-based active learning for rare category detection. In: Advances in Neural Information Processing Systems, pp. 633–640 (2008)
6. Hospedales, T.M., Gong, S., Xiang, T.: Finding rare classes: active learning with generative and discriminative models. IEEE Trans. Knowl. Data Eng. **25**(2), 374–386 (2011)
7. LeCun, Y., Cortes, C.: MNIST handwritten digit database (2010), http://yann.lecun.com/exdb/mnist/
8. Loveland, R., Amdahl, J.: Far point algorithm: active semi-supervised clustering for rare category detection. In: Proceedings of the 3rd International Conference on Vision, Image and Signal Processing. ICVISP 2019. Association for Computing Machinery, New York, NY (2019). https://doi.org/10.1145/3387168.3389117
9. Lowell, D., Lipton, Z.C., Wallace, B.C.: Practical obstacles to deploying active learning. arXiv preprint arXiv:1807.04801 (2018)
10. Pelleg, D., Moore, A.W.: Active learning for anomaly and rare-category detection. In: Advances in Neural Information Processing Systems, pp. 1073–1080 (2005)
11. Svehla, J.: PCKM. https://github.com/open-nlp/active-semi-supervised-clustering (2018)
12. Van Craenendonck, T.: Interactive semi-supervised clustering (2018)
13. Wagstaff, K., Cardie, C., Rogers, S., Schrödl, S., et al.: Constrained k-means clustering with background knowledge. In: ICML, vol. 1, pp. 577–584 (2001)
14. Wagstaff, K., Lu, S.: Efficient active learning in new domains. In: ICML Workshop on Real World Experiment Design and Active Learning (2020)
15. Wagstaff, K.L., Lanza, N.L., Thompson, D.R., Dietterich, T.G., Gilmore, M.S.: Guiding scientific discovery with explanations using DEMUD. In: Proceedings of the Twenty-Seventh AAAI Conference on Artificial Intelligence. AAAI'13, pp. 905–911. AAAI Press (2013)
16. Zhu, X.J.: Semi-supervised learning literature survey. Technical report. University of Wisconsin-Madison Department of Computer Sciences (2005)

Author Index

A
Abel, Marie-Hélène
Abujar, Sheikh
Achchab, Said
Agarwal, Anupam
Ani, Jannatul Ferdous

B
Bensaada, Abdel Kamal
Bhagile, Vaishali D.
Brigui, Iméne

D
Deshmukh, R. R.
Dhillon, Anamika
Dudak, Muhammed Nuri

E
Ekici, Mehmet
Elawady, Iman

G
Gaikwad, Ashrika

H
Herrera, Francisco
Hossain, Syed Akhter

I
Islam, Mirajul
Iyer, Sailesh

J
Jyotiyana, Monika

K
Kalia, Arvind
Kaplan, Noah
Kar, Amit
Kesswani, Nishtha
Kharat, Amit
Khosla, Nitin
Kondal, Sudeep
Kulkarni, Viraj
Kumar, Naveen

L
Lahoud, Christine
Litim, El Mostafa
Loveland, Rohan

M
Martínez-Cámara, Eugenio
Masum, Abu Kaisar Mohammad

N
Namdev, Nischay
Nashed, Nader N.
Nefla, Ons
Nitwane, Rashmi R.

L. Troiano et al. (eds.), *Advances in Deep Learning, Artificial Intelligence and Robotics*, Lecture Notes in Networks and Systems 249
https://doi.org/10.1007/978-3-030-85365-5

O
Ozturk, Meltem
Öztürk, Şaban

P
Pant, Aniruddha
Panwar, Subhash
Purohit, Naresh

R
Redjimi, Kenza
Redjimi, Mohammed
Ria, Nushrat Jahan
Ruiz-Millán, Jose Antonio

S
Sharma, Dharmendra
Sánchez, Francisco Luque

T
Tabik, Siham
Temsamani, Yassine Khallouk

U
Ünal, Yavuz

V
Verma, Gyanendra K.
Viappiani, Paolo
Victoria Luzón, M.
Victorino, Alessandro Correa
Vishwakarma, Monika